The Lithium Legacy

The Lithium Legacy

Ihor Kunasz

Jenny Stanford
Publishing

Published by

Jenny Stanford Publishing Pte. Ltd.
101 Thomson Road
#06-01, United Square
Singapore 307591

Email: editorial@jennystanford.com
Web: www.jennystanford.com

British Library Cataloguing-in-Publication Data
A catalogue record for this book is available from the British Library.

The Lithium Legacy

ISBN 978-981-4968-31-7 (Hardcover)
ISBN 978-1-003-37236-3 (eBook)

To my wife Zenovia for her love, patience, support and encouragement, and who stayed home to raise our wonderful two children Markian (COL. Retired) and Marta (Kunasz Hill), while I trekked over many continents in search of lithium.

Contents

Acknowledgments

My career in lithium would not have been possible if not for the mentoring, the guidance and my relationship with many wonderful friends and professionals and specially my wife who stayed home to raise our children while I searched for lithium all over the world.

Back in 1959, I might not have pursued a career in geology were it not for Dr. Francis Greenough Stehli who became chairman of the geology department at Western Reserve University (today Case-Western) from 1960–1973. He was likely responsible for my graduate assistantship at the Pennsylvania State University.

When I joined Foote Mineral Company, I was lucky to work with the real lithium pioneers, many of whom are no longer with us. I must thank Dr. Wayne Barrett, president of Foote Mineral Company, an exceptionally fine physical chemist, who solved the production of lithium carbonate from complex Nevada and Chilean brines. I thank Mr. Evan P. Comer, then vice-president of Foote Mineral Company who supported me throughout my career and elevated me from Staff to Chief Geologist, and with whom I traveled the world visiting lithium deposits in China, Argentina and the rather eventful visit to the Democratic Republic of the Congo. An avid reader, we shared many hours of interesting and enjoyable discussions during our trips. My sincere thanks to my wonderful secretary Marjorie Abernathy, who organized my disorganization for over 18 years.

To my professional friends, I thank the pioneers of lithium brines, the wonderful people at the Silver Peak operations, John Bassarear, the first General Manager, who appreciated my contribution to the understanding of this virgin brine deposit; Mr. Evans (as I called him), the gruff well field supervisor (who had no use for this freshly-baked university fellow) who came to appreciate my understanding and guidance in locating drilling targets; David Coghlan, the consummate engineering perfectionist, who had no use for poor data (the Hertz Rental Car Company would refuse to lend him a car to Silver Peak, because he always brought back the rental coated with salt). Dave and I spent many years working on the Salar de Atacama brine project. Eugene Dezmelyk, a Ukrainian compatriot from Philadelphia, Foote's director of engineering who designed and built the chemical plants in Nevada, North Carolina and Chile. Dr. Patrick Brown and his staff, Dan Boryta and Bill Fanus, for developing the flow sheet for the complex salar brine and running hundred of brine analyses. I am especially grateful for Rex Bell, my staff geologist, who spent many hours supervising exploration on drill rigs at Silver Peak and who helped me with the initial exploration of the Salar de Atacama. To my Chilean friends, first and foremost I thank Fernando Ide Yager, with whom I did most of the field exploration within and outside the Salar, Felipe Anguita, Manager of the Salar brine project, with whom my family developed a very close friendship and have enjoyed his hospitality in Chile, as well as Eduardo Morales, first manager of the lithium carbonate plant at La Negra (Antofagasta) and Pedro Pavlovic of CORFO with whom I developed a great professional relationship.

I would be remiss not to mention both Phillip Walsh, a brilliant negotiator, whose perfect grasp of Spanish was of invaluable help in our joint-venture negotiations with the Chilean government and Enrique Puga, the Santiago lawyer who walked us through the hoops of Chilean law and also Enrique Artega, the able General Manager of the Sociedad Chilena de Litio Joint venture.

Ihor Kunasz

Introduction

Why write a book about lithium? After all, lithium—the element—is an infant compared to the other metals humans have used. Gold has been used since 6000 BC, copper since 4200 BC, silver since 4000 BC, and iron from 1500 BC. These, either in pure form or admixed with other elements, were visible in surface rocks. Early metallurgists, whether through accident or primitive technology, were able to produce materials useful to society. Other elements, however, were totally unknown until recently, when scientific curiosity prompted researchers to investigate and identify new elements. And so it was with lithium, which was born in 1847. At first a laboratory curiosity, lithium has been part of our society ever since. Although unknown to most people, lithium is used in many practical applications, such as ceramics, greases, electronics, medicine and hopefully, in the future in fusion that would solve the world's energy problems.

Lithium—this unique element—third in weight after hydrogen and helium—does not occur as a pure element like copper, gold, or silver, but is incorporated in various minerals. Lithium was even recently discovered as a component in some brines. In 1966, only one lithium brine deposit (Clayton Valley, Nevada), was known. It is only in the past few decades that it has taken on a life of its own—triggered fundamentally by climate change, which prompted its application and demand in electric cars and storage batteries. In addition to pegmatites, lithium has been discovered today in a plethora of arid closed basins associated with ancient volcanism, which provided the source of lithium. The most important such brine deposit is the Salar de Atacama in Northern Chile.

The author's relationship to lithium began in 1968 at Pennsylvania State University, where as a graduate assistant, he completed a master's thesis on the origin of salt deposition in the Salina Basin of Michigan. This proved to be the gateway to his long career in lithium. Although some lithium had been identified in the brines of Searles Lake California and recovered as a by-product of boron, it had never been identified in other brines. This changed in 1966 when the Leprechaun Mining Company discovered some significant concentrations of lithium in the brines of Clayton Valley—a dry lake located in south central Nevada. Foote Mineral, one of the two major companies which produced lithium from spodumene, acquired the claims but had no expertise in saline deposits. Foote Mineral company contacted the Geology Department of Pennsylvania State University in search of a graduate student who could help with this new project. The only candidate with the appropriate saline background turned out to be the author, who was able to unravel the geological history and the origin of lithium in the Clayton Valley Basin. After spending three summers at the brine operation (1967–1970), he defended a doctoral thesis on the "Geology and Geochemistry of the Lithium Deposit in Clayton Valley, Esmeralda County, Nevada." Kunasz was subsequently hired by Foote Mineral Company to explore for other potential lithium brine deposits, lithium became the author's life's consuming passion. While the literature is full of articles on specific aspects of lithium (chemistry, geology, geochemistry, hydrology, economics), no comprehensive study has been done. This book documents the author's 45 years of a wonderful and exciting association with lithium.

Chapter 1

Lithium Discovery

When discussing or referring to lithium, most publications make casual reference to the discovery of lithium to Johann August Arfwedson from the analysis of a sample from Utoe Island, Sweden. The mineral was collected by José Bonifacio de Andrada e Silva. These were the early 1800s, a time of discovery of many elements by many famous scientists, and so was also for lithium.

Figure 1.1 José Bonifacio de Andrade e Silva.

The Lithium Legacy
Ihor Kunasz

ISBN 978-981-4968-31-7 (Hardcover), 978-1-003-37236-3 (eBook)
www.jennystanford.com

Petalite (lithium aluminum silicate) is a mineral named from the Greek word “petalon” for its perfect basal cleavage. The great Brazilian scientist José Bonifacio de Andrada e Silva (1763–1838) undertook a mineralogical journey through Scandinavia. On his way to Sweden, he studied under Antoine Lavoisier and the Abbé René Just Hauy in France. In 1800, in a letter to mine surveyor Beyer of Schneeberg, he described an infusible laminated mineral from Utoe, Sala, and Finngruva near Kopperberg, which he described as petalite. He returned to Coimbra in Brazil to teach metallurgy. The island of Utoe, off the coast of Sweden, is renowned for its rich iron mines, which has been mentioned in the literature of 1607. It is believed that iron ore was mined as early as the 12th century but poor economics shut the activities down in 1879.[1] Utoe is a typical type locality for four minerals:

- Holmquistite: Discovered in 1913, “Holmqvistit” is unique to the island; it also occurs at the Kings Mountain Foote Mineral Company mine
- Manganotantalite

Figure 1.2 Utoe Island Mine where petalite was collected.

[1]A hundred years before, in 1719, the island was invaded by Russians, and mines, houses, seeds, and cattle were destroyed.

- Petalite
- Spodumene

Samples of those rare minerals are on display in the local mining museum. In 2008, the Swedish Mineralogical Society organized a symposium held at the museum, followed by a two-day trip to the island, where one could still collect samples of petalite and study the lithium-cesium-tantalum (LCT) pegmatites [1].

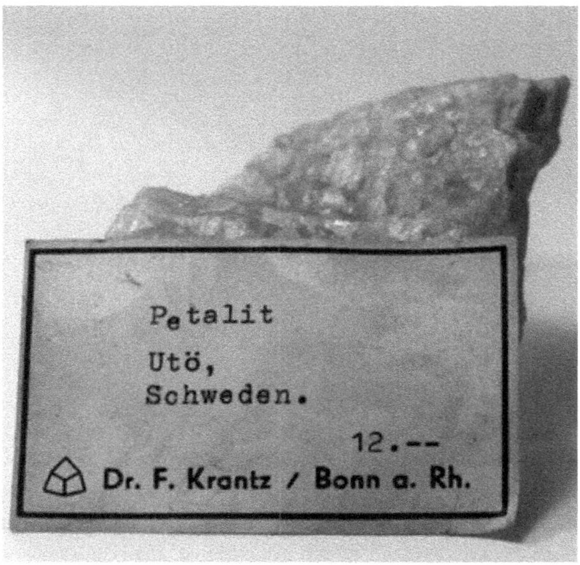

Figure 1.3 Original Petalite specimen from Dr. Krantz collection.

Mineralogists long remained in doubt as to the existence of petalite, which remained in obscurity until 1817 when the mineral was rediscovered by the metallurgist E. T. Svedenstjerna. Petalite had also been reported at the copper mines in Finngrufva near New Kopparberg and the Sala silver mine. Johann August Arfwedson who studied with and worked together with Jons Jacob Berzelius, the renowned Swedish chemist, was asked to analyze the physical and chemical properties of this "unknown" mineral. He discovered the presence of a new element while analyzing the petalite specimen.

Figure 1.4 Johan August Arfwedson.

In his first report [2], Arfwedson states

> The mine of Utoe has long been of great mineralogic importance—several minerals having been discovered there. Investigators, especially in recent years, have tried to discover their chemical composition. However, there are still some minerals, which are familiar enough as regards their external characteristic, have never been subjected to a rigid chemical analysis. Of those specimens, already examined, the analysis showed too little agreement to justify a positive conclusion concerning their composition. It is for these reasons that I have investigated several of them.

The history is somewhat fascinating because, at that time, the chemical analysis of any mineral or compound needed to account for all the elements present. In analyzing the petalite mineral, Arfwedson found that the mineral did not behave chemically according to the rules. Running through a conventional test on 2 grams of petalite, he finished with a 4% loss that he could not account for [3]. His chemical test consisted of

(a) fusing petalite with potassium carbonate,
(b) determination of the silica content, and
(c) precipitation of alumina with ammonium carbonate.

The result of this testing gave the following results:

Compound	Grams	Percentage
Silica	1.564	78.2
Alumina	0.356	17.8
Loss	0.080	4.0

Unsatisfied with the results, he performed three additional tests with similar results. He guessed that the lacking compound was sodium sulfate and concluded that it explained the loss. The last analysis, like the previous ones, gave the following—silica, 79.85%; alumina, 17.30%; sulfate, 17.75%—which gave an incorrect total.

After several attempts, Arfwedson concluded that, "Finally, after having studied most closely the sulfate in question, I found that it contained a particular fixed alkali, the nature of which was not yet known. Professor Berzelius proposed the name 'lithion' for it (from the Greek[2] word 'lapideus,' because this alkali was first found in the mineral kingdom." The name was later standardized as "lithium." The three sets of results give the following molar ratio ranges:

$$SiO_2: Al_2O_3:Li_2O\ 8.22–8.23:1.0–1.05:1.00–1.08,$$

which compare favorably with the theoretical ratio of 8:1:1. The present-day analysis of petalite corresponds closely to SiO_2, 76.16%; Al_2O_3, 17.24%; Li_2O, 4.49%. Arfwedson later showed that this same element was present in spodumene and lepidolite. Arfwedson was successful in producing small amounts of lithium metal. Some reference works suggested Brandy isolated lithium in 1822, but it was in 1818 that Sir Humphrey Davy obtained a minute amount of lithium metal by passing a current through fused lithium carbonate in a platinum capsule. In 1841, during the last year of his life, Arfwedson was awarded a gold medal to honor his discovery of lithium by the Swedish Academy of Science.

It is interesting to note that petalite was thought to be an exclusively Swedish mineral. However, Gerald Troost, a Dutch and a student of Abbé Hauy, the founder of the Philadelphia Academy of Sciences, obtained samples of petalite (which he considered to be tremolite) from the North Shore of Lake Ontario, Canada. In the

[2]*Author's note*: Arfwedson incorrectly referred to the name as Greek when in fact it is Latin for Stony.

United States, Thomas Nuttall sent spodumene samples from Sterling Massachusetts. Spodumene had never been found there before. In 1818, Christian Gmelin was the first to observe that lithium salts give a bright red color in flame. However, both Arfwedson and Gmelin tried and failed to isolate the element from its salt. Davy probably received a lithium salt from Berzelius shortly after its preparation by Arfwedson. The *Annals of Philosophy* had already carried a report in the May issue that Davy had succeeded in isolating the metallic element. In 1821, William Thomas Brandy isolated the element by performing electrolysis on lithium oxide. In 1855, Robert Bunsen and Augustus Matthiessen produced large amounts of the metal through the electrolysis of lithium chloride. However, it was not until 1923 that the application of this procedure led to the invention of the commercial production process of lithium metal by the German company Metallgesellschaft AG through electrolysis of a liquid mixture of lithium chloride and potassium chloride. Commercial production of lithium carbonate and finally lithium chloride from zinnwaldite, a mineral mined in eastern Germany started in 1923 at Metallgesellschaft AG Langelsheim plant [4].

References

1. Grew, E. S., Jonsson, E., and Langhof, J., 2018, Lithium: 200 years, symposium and field trip June 14–16, 2018, *Elements*, **14**(4), 284–284.
2. Gentieu, N. P., 1957, *Foote Prints,* **29**(2), 18–25.
3. Weeks, M. E., 1933, Discovery of the elements, *Journal of Chemical Education*, **10**(11), 494–503.
4. Metallgesellschaft, 1963, *Review of Activities* (6), 27–31.

Chapter 2

Dr. Albert E. Foote (1846–1895)

The present state of the lithium industry cannot be discussed without mentioning the avid mineral collector who ultimately caused lithium to be the buzzword of the 21st century.

Figure 2.1 Dr. Albert E. Foote.

Albert Edward Foote, one of America's most famous early minerals dealers, was born on February 4, 1846, in Hamilton, Madison County, New York, as son of Edward Warren Foote and

The Lithium Legacy
Ihor Kunasz

ISBN 978-981-4968-31-7 (Hardcover), 978-1-003-37236-3 (eBook)
www.jennystanford.com

Phoebe Steere. He graduated from Courtland Academy in Homer, New York, where he first became interested in mineralogy through the influence of Dr. Caleb Green and began collecting minerals in 1862. He was a student of Prof. Walcott Gibbs at Cambridge and Prof. Hoffmann in Berlin. Oddly enough, Foote obtained his medical degree in 1867 from the University of Michigan in Ann Arbor, USA. In 1870, after teaching natural history rather than medicine for three years at Ann Arbor, he took the position of assistant professor of chemistry and mineralogy at the Iowa State Agricultural College and was promoted to full professor in 1871 [1].

He married Augusta Matthews in January of 1872, in Knoxville, Iowa. In December of that year, they had a son, Warren Mathews Foote, who was destined, 23 years later, to take over his father's business. During this time he started collecting minerals from the Iron Range of Michigan. His problem, which propelled him into becoming one of the famous mineral collectors, was that he could not leave behind anything that looked like a mineral. He thus amassed a large inventory of barrels and boxes full of minerals. He then moved to Iowa, where he taught chemistry and mineralogy at the Iowa State College, USA. There he had the opportunity to travel to Europe to visit some of the best analytical laboratories, and, upon returning, set up what was to become one of the top analytical laboratories in the United States.

Figure 2.2 The A. E. Foote Pavilion at the Philadelphia Centennial Exhibition.

The year 1876 was pivotal for Dr. Foote—it was the year of the Centennial Exposition in Philadelphia. He resigned from his position at the Iowa State College, packed his collection (not only of minerals but also medical books, shells, fossils, eggs, and all objects of natural history) and rented a cheap building on North 4th Street. His exhibit was so impressive that he was awarded first prize for scientific interest and beauty of form and color. He became instantly famous and, by 1889, he had collected over 700,000 mineral specimens.

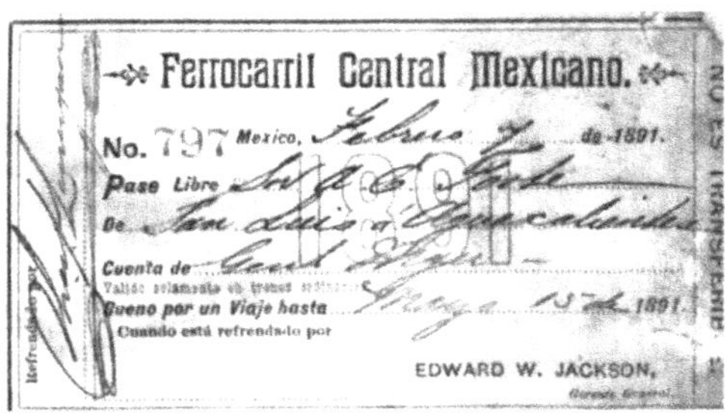

Figure 2.3 A. E. Foote train ticket to Mexico.

Sadly, Dr. Foote was not a well man; he suffered from tuberculosis and would escape the unpleasant winters of Philadelphia for warmer climes. However, minerals were always on his mind, and he endeavored to augment his supply of minerals during his travels. For example, he would stop at Joplin, Missouri, with a pocket full of cheap minerals that he would use to entice would-be collectors and invited them to his hotel room where he would show and sell some fine specimens. He also would use the fact that he was a medical doctor to contact the local physician who knew the local miners who had specimens in their houses and proceeded to purchase all they had. He would then ship from a dozen to twenty boxes to his Philadelphia store. He traveled to Pike's Peak, famous for the amazonite and smoky quartz specimens; from there he would visit the famous Red Cloud mine in New Mexico, where the best wulfenite crystals have been collected. An astute businessman, he would identify which minerals would go to famous collectors such as Bement and Vaux. With winter setting in, he would travel to Mexico where he would send some spectacular specimens back to Philadelphia for sale. He traveled by train, and his tickets would reflect the fact that Dr. A. E. Foote traveled with several boxes of minerals. He was a contemporary of Ward (of the famous Ward catalog), who visited him after Foote's return from a trip to Sardinia, where he had collected many fine anglesite and phosgenite specimens. Foote met his friend and sold him what Ward believed were the best of the lot. However, six months later, when Ward visited the Foote's Philadelphia store, he saw the best samples. He called Foote a scoundrel and both laughed [2].

Figure 2.4 A. E. Foote Amazonite specimen—Pike's Peak, Colorado.

Foote was a prolific collector of fine minerals and published many catalogs. He even offered an artistic assemblage of minerals for the handsome sum of $350. Most museum mineral collections included specimens supplied by Dr. Foote. After his death at the young age of 49 (November 10, 1895), many testimonials (George F. Kunz after whom kunzite, the purple gem spodumene, was named and Edward S. Dana, who had compiled the extensive *System of Mineralogy* and after which Dr. Foote cross-referenced his own 1909 *Complete Mineral Catalogue*) explained his invaluable contribution to the field of mineralogy. His very first *Catalogue of Minerals and Tables of the Specimens* was published in July 1880. The latest edition of 240 pages, which was offered free to teachers, could be bought for $0.25 [2].

Figure 2.5 A. E. Foote artistic mineral composition.

Dr. Foote received numerous awards and medals for his exceptional mineral displays in Philadelphia (1876), Cincinnati (1881), New Orleans (1884–1885), Louisville (1886), London (1887), and Paris (1889–1890).

Figure 2.6 A. E. Foote Award Medals for mineral displays.

His son, Warren M. Foote, had written[1] to his mother in 1900 that Dr. Foote had agencies in Buenos Aires, Constantinople, Singapore,

[1]From the original personal correspondence of Warren M. Foote available.

Yokohama, and Melbourne in addition to their main houses, located at 24, Rue du Champ de Mars in Paris as well as in Berlin.

Figure 2.7 Foote Mineral Company letterhead, Paris, France.

In another letter to his mother, he indicated that the business is not as lucrative as he had expected and that, if it does not improve, he would have to close shop. Then what happened to make the business of Foote survive and ultimately become such an important company?

Warren had been collecting minerals with his father and eventually took over the business. But it was not what his father would have wished. Warren made some significant changes. A major event changed the course of the Foote Mineral Company from being a purveyor of fine mineral specimens to a supplier of large quantities of specialized minerals. This came about when Warren received a plea from his Paris representative to bring a display of minerals to the Paris exhibition. He was less than enthusiastic because to assemble and ship such a display would have been quite expensive. Finally, he acceded and sailed to Paris, and this was to become a very auspicious journey—one that changed the future of the Foote Mineral Company. The German company Siemens and Halske had come to the Paris exhibition in search of a supplier for tantalum ore, which was virtually unknown in 1900. The Germans, improving on Thomas Edison's carbon filament incandescent bulb, inquired about one of the specimens exhibited; the specimen was columbite (a complex iron-manganese-magnesium-niobium-tantalum oxide ore).

The Germans wanted to know if Foote could provide large quantities of the mineral. Warren Foote knew a great deal about tantalum ores and where to find them. Over the next six years, he supplied Siemens and Halske minerals valued at over $200,000—a welcome infusion into the company's business. It is reported that Warren promptly had all the fine columbo-tantalite specimens in the Philadelphia store crushed and shipped to Siemens. This apparently caused such an uproar that many workers resigned.

Tantalum metal has a melting temperature of 5,463°F and replaced the earlier carbon filament lamps. The tantalum lamps were a great success and over 103,000,000 lamps were sold from 1905 to 1911. Later the lamps were replaced by tungsten lamps [3].

Delighted with his success, Warren announced that the company would, henceforth, be called the Foote Mineral Company. Yet, lithium was not one of the main commodities supplied by Warren because zirconium took priority since Foote had invested in zirconia refractories.

Figure 2.8 Foote Mineral Company official letterhead—Philadelphia.

So how did Foote Mineral Company enter the lithium business? As early as 1909, Thomas Edison had successfully developed an alkaline storage battery.

However, the United States was not a producer of lithium chemicals, because no significant lithium ore deposits had yet been discovered, save some minor one in the Black Hills of South Dakota. Most production of lithium chemicals came from Europe. When a large glass manufacturer approached Foote for a large quantity of lithium carbonate, Gordon Chambers, an early associate of Warren Foote, realized that large quantities of lithium minerals would be required.

A mapping report of the Kings Mountain belt by A. Keith [4] indicated the presence of spodumene but was long-forgotten. Gordon Chambers bought it and obtained a ton of spodumene for gravity concentration. Gravity concentration, which relies on the difference in specific gravity, worked well for precious metals but failed to concentrate spodumene. The United States Bureau of Mines (USBM), however, successfully recovered spodumene by the flotation process, but the Foote Mineral Company abandoned the process because of the high capital required. Then, early in World War II, there was a great requirement for lubricating greases. Foote obtained part of the contract, while Solvay Process Company (Allied Chemical) obtained leases on the best part of the Kings Mountain spodumene belt. Foote was producing lithium at its Exton PA plant that it had bought from the U. S. government. It acquired the Kings Mountain mill that Solvay had decided to sell. With an assured raw material source, Warren approved a new chemical processing plant at Sunbright, Virginia, using spodumene concentrates from Kings Mountain. Finally, Foote Mineral Company decided to abandon the zirconium business and concentrate its effort in the production of lithium for the ceramic and greases industry. Tony Bliss, who later became Foote's chairman, had prophetically said, "I see a rainbow with two pots of gold: one is lithium in ceramic products and the other is lithium-based grease." Foote Mineral was on its way to become the premier lithium company.

References

1. Kraus, E., 1959, Albert E. Foote, the naturalist: a Michigan Alumnus. *Quarterly Review*, **64**(21), 341–347.
2. Foote Mineral Company, Laboratory material, and individual specimens of the common and more important minerals, *Catalog of Elementary Collections.*
3. Toothaker, C. R., 1951, The days of A. E. Foote, *Rocks and Minerals*, **26**(9-10), 460–463.
4. Keith, A., and Sterrett, D. B., 1931, Gaffney-Kings Mountain Folio, South Carolina–North Carolina: Folios of the Geologic Atlas. USGS Numbered Series report, No. 222, doi. 10.3133/gf222.

Chapter 3

Lithium: The Element

Lithium is a soluble alkali element. Because its ionic radius is small (0.78 Å), it behaves more like magnesium (0.72 Å) than the alkalis. Li^+ tends to substitute for Al^{3+}, Fe^{2+}, and especially for Mg^{2+}.

It is a metal and the lightest solid material in the universe. It is the third element in the periodic table preceded in the periodic table by only two gases—helium and hydrogen. It weighs less than the glamorous lightweights of the industry—aluminum and magnesium. With a specific gravity of 0.534, it will float in water like an ice cube. This buoyancy exacts a drastic penalty, however, because, as it floats it disintegrates, much like an Alka–Seltzer tablet and ignites violently. It can be sliced with a sharp knife nearly as easily as a cheese. When first cut, it gleams brightly like polished silver, but, if left exposed to air, its surface quickly tarnishes to a dull black patina.

The name of the third element is lithium. It is related to sodium and potassium, but often it acts like a stranger to its chemical family. Its behavior is frequently atypical of the alkali metals, the group in the periodic table to which it belongs. Its unorthodoxy has been the secret of its success. From chemical and physical properties that are unique have come innovations that are now established as standard industrial applications. From its early discovery, its uncommon properties implied the promise of developments yet undreamed of. In a single atom of lithium, three electrons whirl in space around a

The Lithium Legacy
Ihor Kunasz

ISBN 978-981-4968-31-7 (Hardcover), 978-1-003-37236-3 (eBook)
www.jennystanford.com

nucleus at an incredible speed. To visualize this microcosm, imagine two electrons pursuing inner orbits in the manner of Earth and Venus in the solar system. Then further imagine a solitary valence electron that moves in its remote orbit as Pluto circles the sun—3.5 billion miles. But it is this outer electron that to a large extent determines the unique chemical behavior of lithium and, therefore, enhances its status as a material of extraordinary potential.

The ease with which it becomes involved with another atom, it can readily change a neutral lithium atom into a positive lithium cation. This transforms the nuclear character of the element—the lithium cation has an extremely small volume and a high positive electrical charge. With only two electrons, the positively charged nucleus exerts powerful forces on other atoms. Therefore, lithium forms very stable compounds with Group II elements, beryllium, magnesium, calcium, strontium, and barium, with the release of large amounts of energy. With elements like oxygen or fluorine, the energy released per unit weight is among the highest of any chemical reaction.

A small amount of lithium metal was produced in 1818 by Sir Humphrey Davy. One of the first practical application of lithium grew out of an investigation by Thomas Edison. He became interested in the use of lithium hydroxide to add "pep" into his alkaline storage battery. It effects, which he admitted he could not explain, is recorded in a patent filed on May 10, 1907, in which Edison explained that adding 2 g of lithium hydroxide to every 100 cc of electrolyte solution caused his battery's capacity to spike by 10% and extended the amount of time the battery could hold a charge by a "remarkable" amount.

In Germany, too, the possibilities of unexpected behavior were not lost on astute scientists when the blockade imposed during World War One interrupted the country's supply of tin German metallurgists alloyed lead and lithium experimentally as an ersatz metal for the bearing parts of railroad car trucks. Although the results seem not to have been overwhelming, this early and daring venture does assume special importance as a pioneering project and it probably stimulated later and more successful investigations into alloys of lithium with other metal such as aluminum (which was purported to be a replacement for all the steel). For the most part, large scale industrial development of the third element remained a dream of the future and a rather vague one at that, until the comparatively recent era of the early 40s. The global warfare of just two decades ago brought with it a desperate urgency to find substitutes for all the traditional materials that suddenly seemed to be in short supply. It was for this reason that lithium underwent a belated stimulus that began to reveal some of its unique values and greatly increased the requirements for its production.

For almost a century, lithium languished in obscurity as a laboratory curiosity. Its only value resided in its supposed healing powers. Spas in Europe and in this country known to contain this therapeutic element attracted a steady clientele of rich hypochondriacs and chronic sufferers of nagging ailments such as gravel (kidney stones) and gout. Scientists reported in 1994 that controlled experiments of males and females bathing for two weeks in spas did not show any increase in serum lithium [1].

During the 1950s, a special event strongly influenced the course of the lithium destiny. It took the form of special interest by the Atomic Energy Commission in one of the isotopes (forms of the same element that have closely related properties but slightly different atomic weights)—lithium^6—as an indispensable part of its application in nuclear power. It was the foundation of the first hydrogen bomb—a thermonuclear reaction, which if contained could solve the energy requirement of the planet. The result of this massive demand was a further improvement in production techniques and a more precise evaluation of lithium ore deposits as a realistic inventory of the long-term supply potential of the third element.

The reactive isotope, Li^6, which comprises only 7.52% of the natural element creates an enormous amount of energy when bombarded with electrons according to the following formula:

$$^{6}Li + n \rightarrow T + {}^{4}He + 4.8\ MeV,$$

which then can lead to the following fusion reaction:

$$D + T \rightarrow {}^{4}HE + n + 17.6\ MeV$$

A huge amount of energy was released upon the explosion of the bomb; however, containment was another issue if thermonuclear power were to become a practical alternative to fission reactions used in the present nuclear plants. That is because the initial reaction temperature is that of a solar-core temperature of around 15 million Kelvin or 27 million degrees F. No known containment material can stand such temperatures. Therefore, reactors need to contain the plasma until the initial reaction has taken place. Existing reactors such as the Tokamak, ITER (International Thermonuclear Experimental Reactor) use magnetic fields to contain the plasma.

In 1957, Dr. Milton Rothman of the Bartol Research Foundation of the Franklin Institute published an article titled "The short-lived radioactive isotopes" in which he identified a great number of lithium isotopes [2].

THE ISOTOPES OF LITHIUM

Li^4 Existence questionable.
Li^5 Lifetime approximately 10^{-21} seconds.
Li^6 7.52% of natural lithium.
Li^7 92.48% of natural lithium.
Li^{7m} Half-life 5.3×10^{-14} seconds.
Li^8 Half-life 0.841 seconds.

$$Li^8 \xrightarrow{\beta^-} Be^8 \longrightarrow 2\,He^4$$

Li^9 Half-life 0.168 seconds.

$$Li^9 \xrightarrow{\beta^-} Be^{9*} \longrightarrow Be^8 + n$$

Figure 3.1 The isotopes of lithium.

The isotopes have been used by several researchers to interpret the origin of various geological environments. For example, the Li isotopic signature of lunar basalts is like the isotopic signature of the Earth's mantle. This indicates core formation, volatile loss, and the presence of the crust and hydrosphere have not significantly influenced the Earth and the Moon, which are already differentiated planetary bodies [3]. Isotopes from lunar samples and Martian meteorites and terrestrial basalts *and* peridotites show that there is no difference between Mars, Vesta, the Moon, and the Earth [4].

Lithium is a very mobile element that incorporates itself into many minerals and brines. Lithium and magnesium substitute for aluminum in the octahedral structure of mica and minerals such as hectorite.

Its unique properties resulted in the various applications. Lightweight lithium alloys used in the manufacture of aircraft have the potential to reduce weight by as much as 10%. Li/Al alloys contain up to 7.5% lithium and Li/Mg alloys can contain up to 13% lithium, while most other lithium alloys contain 2% to 3% lithium [5].

The third element, after lying dormant for almost a century after its discovery, had come to age, especially in the late 20th century, when it became the main source of power in most consumer electronics such as cell phones and laptops. The push to convert the internal combustion engine to electric engines is driving the demand for lithium batteries because of their high energy per unit mass and high power-to-weight ratio. Lithium metal is the basic ingredient of the battery [6].

Converting lithium into metal is done in an electrolytic cell using lithium chloride. The lithium chloride is mixed with potassium chloride in a ratio of 55% to 45% to produce a molten eutectic electrolyte. Potassium chloride is added to increase the conductivity of the lithium while lowering the fusion temperature.

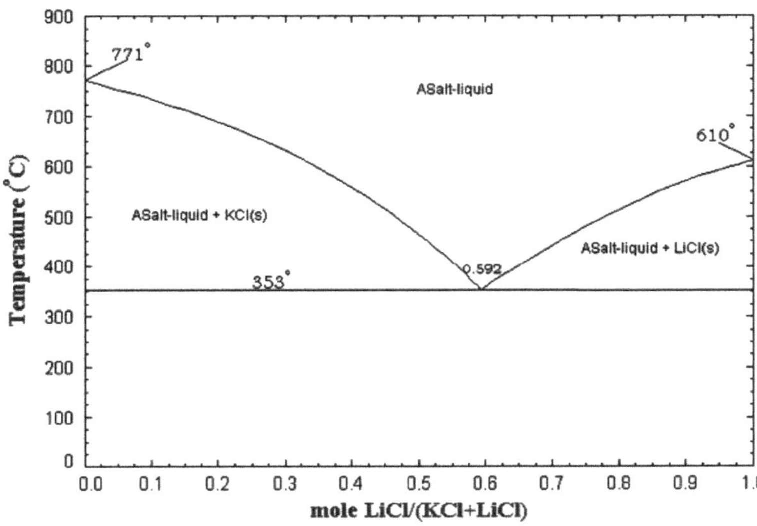

Figure 3.2 Lithium metal eutectic.

Lithium-ion batteries (LIBs), while first commercially developed for portable electronics, are now ubiquitous in daily life in increasingly diverse applications, including electric cars, power tools, medical devices, smartwatches, drones, satellites, and utility-scale storage.

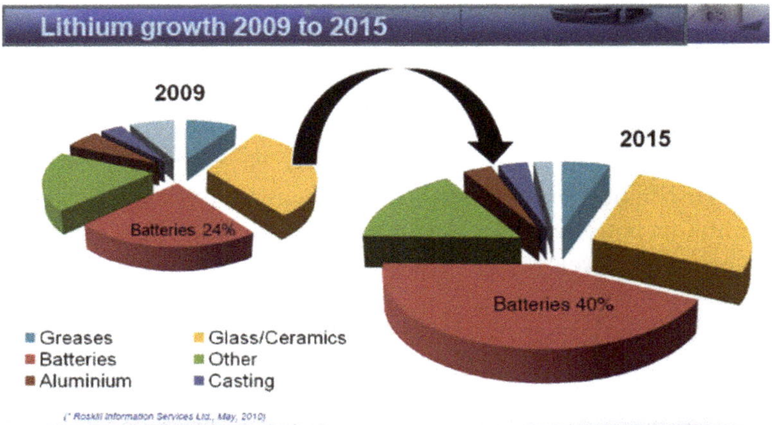

Figure 3.3 Lithium applications—2009–2015.

The first challenge for researchers is to reduce the amounts of metals that need to be mined for electric vehicle batteries.

References

1. McCarthy, J. D., Carter, S. P., Fletcher, M. J., and Reape, M. J., 1994, Study of lithium absorption by users of spas treated with lithium ion, *Human & Experimental Toxicology*, **3**(5), 315–319.
2. Rothman, M., 1957, The short-lived radioactive isotopes, *Foote Prints*, **29**(1), 3–14.
3. Makishima, A., 2017, The giant impact made the present Earth–Moon system, in *Origins of the Earth, Moon and Life*.
4. Halliday, A. N., 2014, Planets, asteroids, comets and the solar system, in *Treatise on Geochemistry*, 2nd Ed.
5. James, R. S., 1990, Aluminum–lithium alloys, in *ASM Handbook*, Vol. 2, *Properties and Selection: Non-Ferrous Alloys and Special Purpose Material*.
6. U. S. Department of Energy: Energy efficiency and renewable energy. Batteries for hybrid and plug-in electric vehicles. Alternative Fuels Data Center.

Chapter 4

Lithium Minerals

Lithium was identified only in 1817 in a petalite specimen from Utoe Island in Sweden. While many authors (this author included) have quoted that 145 lithium-bearing minerals have been identified, the source of this data is unknown. MINDAT [1] reports that there are 95 valid species containing essential lithium (the concentration of lithium is, however, not specified). However, a recent publication has reported the number of lithium-bearing minerals by using a rigorous systematic approach to identify the actual number. In a 2019 publication of the *European Journal of Mineralogy* [2], Grew identified only a total of 118 lithium minerals (and 120 beryllium and 296 boron minerals). Present crystal concentrations are generally too low for lithium and beryllium minerals and require 1–2 orders of enrichment by partial melting and mobilization of fluids, which then lead to high enough enrichment to produce pegmatites (to be discussed in subsequent chapters). Grew concluded that unique conditions at a handful of localities produced a diverse suite of lithium minerals rarely replicated elsewhere. The pace of new lithium-bearing minerals showed an exponential increase since its early identification in 1847 (see Appendix).

The Lithium Legacy
Ihor Kunasz

ISBN 978-981-4968-31-7 (Hardcover), 978-1-003-37236-3 (eBook)
www.jennystanford.com

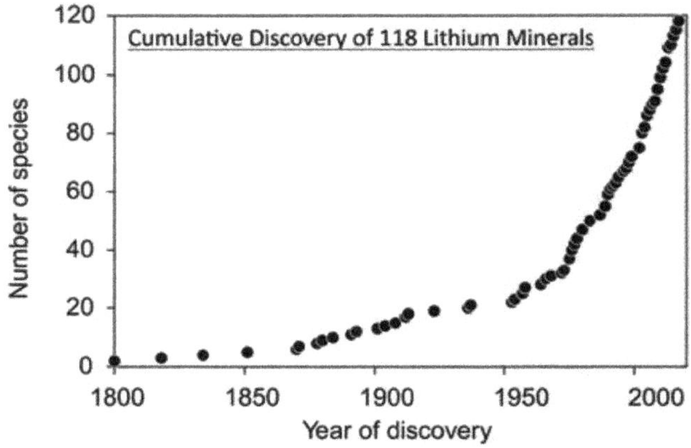

Figure 4.1 Rate of lithium mineral discoveries.

4.1 Age of Formation of Li Minerals

As in the case of beryllium and boron minerals, dating the formation of Li minerals can be problematic, because none of the Li minerals can be dated directly, and thus the ages of Li minerals have been inferred from the ages obtained from associated minerals. For example, 31% of the oldest occurrences are pegmatites that have been dated by U–Pb isotopes in associated minerals in the pegmatite; this represents half of the Li minerals occurring in pegmatites, and thus, in principle, there is potential for more precise dating of Li minerals. Alternatively, Li minerals were dated based on related geologic information, such as an age for the deposit in which these minerals are found. Researchers have been mindful in considering whether Li minerals are younger than the dated mineral or deposit, and fortunately, cases in which Li minerals are significantly younger are fewer than in the case of B minerals. Researchers were unable to find reliable ages for lithiophorite and swinefordite, as both minerals are supergene and considerably younger than the rocks with which they are associated. Researchers did not attempt to get a maximum age for hectorite. Confirmed hectorite is found largely in Tertiary deposits [1] and appears to be supergene where it occurs in pre-Tertiary pegmatites.

Although a great number of lithium minerals have been identified, only spodumene, lepidolite, amblygonite, and eucryptite have enough significant lithium concentrations to have been commercial sources of lithium. Today, spodumene is the principal lithium ore although attempts are being made to resurrect petalite mining from Zimbabwe.

4.2 Spodumene ($LiAlSi_2O_6$)

Spodumene is an alumino-silicate, which is a monoclinic member of the pyroxene group. It has a very pronounced cleavage plane along the 110 direction which results in typical lath shaped particles upon breaking.

In most pegmatites, the dimensions of the spodumene crystals tend to be relatively small (centimeters) but in some of the deposits of the Black Hills of South Dakota, spodumene logs reach lengths of 40 feet and widths of 3.5 feet (note the man standing in Fig. 4.2). Large spodumene crystals occur also in Manono and Kitotolo pegmatites of the Democratic Republic of the Congo.

Figure 4.2 Giant spodumene Logs in Etta Mine—South Dakota.

The color of spodumene is variable, being nearly white in low-iron varieties to dark green in iron-rich crystals. Spodumene undergoes pseudo-morphic alteration to a variety of minerals; Norton and Schlegel (1955) have described spodumene replacement by quartz, albite, perthite, muscovite, beryl, amblygonite, apatite, and tourmaline [3]. Weathering commonly alters spodumene to kaolinite and montmorillonite.

Figure 4.3 Gem quality spodumene crystals—Afghanistan.

When clear, spodumene is considered a semi-precious gemstone. Three varieties are known: *Hiddenite,* the green variety from Alexander County, North Carolina, first discovered in Brazil; *triphane,* the yellow variety also from Alexander County; and *kunzite,* the blue variety from Afghanistan as well as the lilac-colored variety from the Pala District, California. After the mining of lepidolite on Hiriart Hill, about a mile and a half east of the village of Pala, a chalky mass containing very large quartz crystals were obtained. While removing the quartz crystals, beautiful lilac-colored crystals were discovered. Unable to identify the mineral, specimens were sent to Tiffany & Company in New York, where Dr. George Kunz identified them as spodumene. He had previously found such small pieces at Branchville, Connecticut. Charles Baskerville, a chemistry professor at the University of North Carolina, named the mineral kunzite, in honor of Dr. Kunz.

The spodumene crystals from the Nuristan region of Afghanistan are among the finest examples of kunzite ever found. The transparent, gem-quality spodumene crystals from Nuristan come in a wide range of colors—purple and pink, as well as blue, green, and yellow. Some can be a meter long and can make very attractive jewels.

Figure 4.4 Cut Kunzite gemstone.

Spodumene constitutes the most abundant commercial source of lithium. Theoretically, the spodumene crystal contains 3.7% Li, but the actual concentrations may vary from 1.35% to 3.56% Li, probably as a result of sodium or potassium substitution or because of weathering.

In the Tanco Mine in Manitoba, Canada, a mineral called SQI (spodumene-quartz intergrowth) has been reported. Because SQI has a lithium concentration of 4.1% and a very low iron content, it has been interpreted as an exsolution of petalite [4].

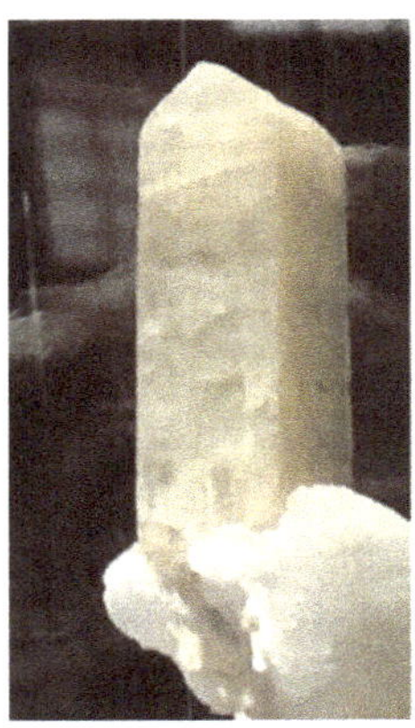

Figure 4.5 Rare terminated spodumene crystals—Foote mine, North Carolina.

Spodumene is mined from two types of pegmatites: unzoned and zoned.

In unzoned pegmatites, such as those in the Tin–Spodumene Belt of North Carolina, the spodumene comprises about 20% by volume and is distributed evenly across the pegmatite body resulting in a lithium concentration of about 0.7% Li. Terminated crystals are rarely found, such as the one I collected at the Foote Mineral Company mine.

In zoned pegmatites such as Tanco, Canada, and Greenbushes in Australia, the spodumene is concentrated in specific zones that tend to be quite rich in lithium (and low in iron—critical for direct ceramic formulations). At Greenbushes, some pegmatite ore zones return upward of 2.3% Li.

Figure 4.6 Iron-rich spodumene in unzoned pegmatite.

Spodumene in unzoned pegmatites tends to have a high iron content compared to spodumene in zoned pegmatites. Three distinct paragenetic types of spodumene can be identified in pegmatites:

(1) Spodumene phenocrysts occur in essentially unzoned pegmatite of the "Kings Mountain" type. These laths are usually less than a foot long, commonly are greenish and contain essential Fe^{3+} in substitution for Al^{3+}, usually in the range of 0.6–0.9% Fe_2O_3. The iron must be removed in order to satisfy the low-iron criteria for ceramic grade applications. As spodumene occurs as a monoclinic dense alpha crystal, the iron is not reachable. However, decrepitation at a temperature of 900°C leads to a lattice expansion of about 27%, converting it to the tetragonal beta form, allowing the acid to attack the iron. At a reaction time of 10 minutes, 78.8% of the iron is liberated [5].

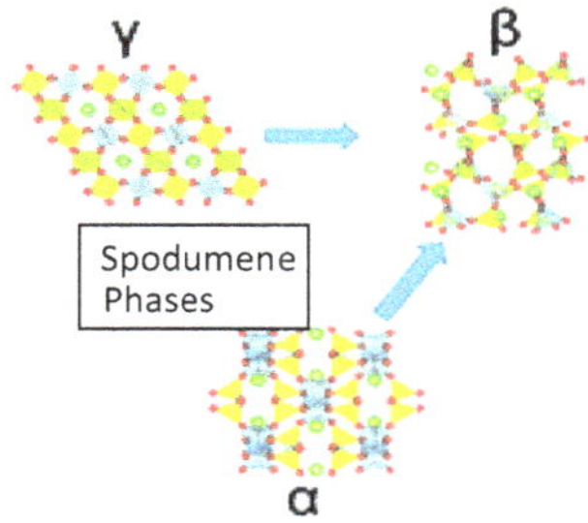

Figure 4.7 Spodumene phase conversion during heating.

(2) Zonal spodumene occurs commonly as large laths in intermediate zones and cores of well-zoned pegmatites. This spodumene is low in iron (Fe_2O_3 = 0.01–0.03%) and it may contain significant manganese (*x*.0–0.*x*%). Whereas most of it is white, some is pink or lilac (*kunzite*) also.

(3) Secondary spodumene, produced by the isochemical decomposition of petalite: petalite + spodumene + 2 quartz, according to the reaction [4]:

$$LiAlSi_4O_{10} \Rightarrow LiAlSi_2O_6 + 2SiO_2$$

This spodumene is relatively fine-grained; the aggregate commonly retains the crystal form and basal cleavage of the parent petalite. It is white and has very low iron (Fe_2O_3 = 0.007–0.03%), inheriting the low iron content of the parent petalite. At the Tanco Mine, the term SQI has been applied to this type of spodumene. At Greenbushes, this appears to be the case as well. The lithium zones in the main pegmatite generally contain coarse-grained euhedral spodumene intergrowth with quartz [6]. This pegmatite is not only extremely rich in lithium oxide, having more than 5%, but also has a very low iron content. The concentrate grading up to 7.6% Li_2O and a concentration of 0.07% Fe_2O_3 can be used directly in ceramic applications. An additional lower grade of 6.4% with 0.4% Fe_2O_3 is shipped to China for chemical conversion.

4.3 Lepidolite [$K(Li,Al,Rb)_3(Al,Si)_4O_{10}(F,OH)_2$]

Lepidolite is a phyllosilicate belonging to the trioctahedral group of micas. The chemical variability expressed in the formula stems from a structural complexity attributed to a three-part series consisting of polylithionite, lepidolite, and trilithionite. All three minerals share similar properties because they have lithium and aluminum (in varying ratios) in their chemical formulas [1].

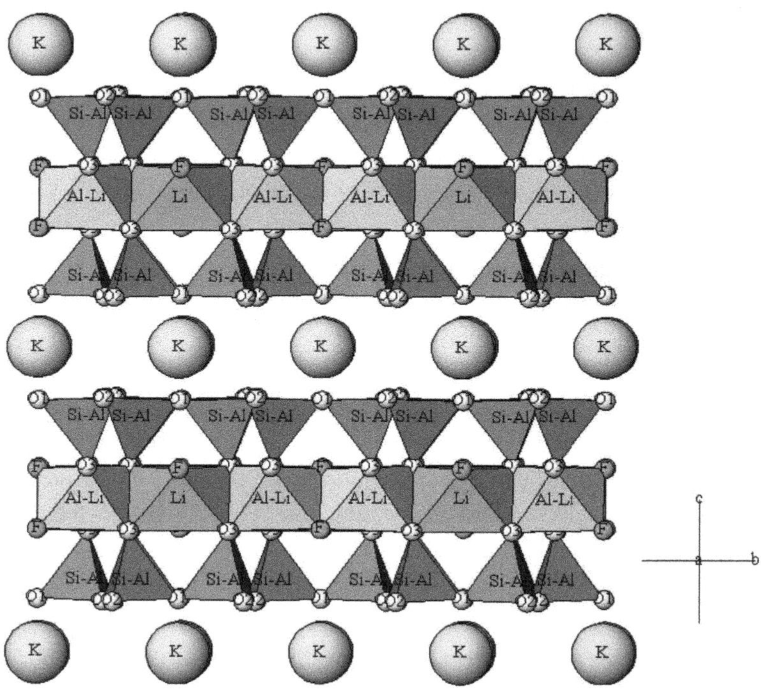

Figure 4.8 Lepidolite structure.

Lepidolite was named in 1792 by Martin Klaproth from the Greek words lepidos for "scale" and lithos for "stone." Abbe Nicolaus Poda of Neuhaus, a Jesuit scholar was presumably the discoverer of lepidolite (or lilatite as he called it). The first published account of lepidolite is Baron von Born' description of a sample from the estate of Count Nepomuk von Mitrowsky. The sample came from the Rožná pegmatite, Žďár nad Sázavou, Vysočina Region, Moravia, Czech Republic [7].

The pink to purple color typical of lithian muscovite and lepidolite is usually attributed to manganese rather than lithium.

Deer, Howie, and Zussman (1962) suggested that there is a continuous series between muscovite with a 2M1 structure to lepidolite with 1M, 2M2, and 3T structures [8]. The structural transition takes place when the lithium oxide content in the mica reaches 1.53%. While lepidolite occurs in small crystals, muscovite mica reaches very large sheets in some pegmatites. Muscovite from Indian pegmatites was imported to be used in front of fireplaces and was known as Eisen glass. Sinkankas wrote "The world's record [muscovite, $KAl_2(AlSi_3O_{10})(OH)$] for size is held by a single crystal from the Inikurti Mine, Nellore, India, which measured 15 feet (4.57 m) in length and 10 feet (3.05 m) in diameter and delivered a total of 85 tons (77,111 kg) of muscovite" [9].

Figure 4.9 Alteration of muscovite to lepidolite.

The crystal forms of lepidolite include thick books with intergrowths, botryoidal growths, and very fine grains.

The lithium concentration in lepidolite ranges from 1.53% Li to a possible theoretical maximum of 3.6%. In commercial deposits, the concentrations are more normally 1.4% to 1.9% Li. In addition to lithium, lepidolites also carry substantial concentrations of rubidium and cesium [8]. Lepidolite is a secondary source of lithium. It is a member of the polylithionite–trilithionite series. It is associated with other lithium-bearing minerals like spodumene in pegmatite bodies.

It is one of the major sources of the rare alkali metals rubidium and cesium. In 1861, Robert Bunsen and Gustav Kirchhoff extracted 150 kg of lepidolite and yielded a few grams of rubidium salts for analysis, and thus discovered the new element rubidium [10].

Deposits of lepidolite are in Zimbabwe (Bikita), Namibia (Karibib), Canada (Bernic Lake, Manitoba), Brazil (Minas Gerais), Portugal, Spain, and the People's Republic of China.

Besides being a commercial lithium mineral used in the manufacture of lithium chemicals, lepidolite had a far more important role in the fusion reaction which resulted in the creation of the hydrogen bomb in the 1950s. The lepidolite used for this purpose was mined in Zimbabwe and is basically exhausted.

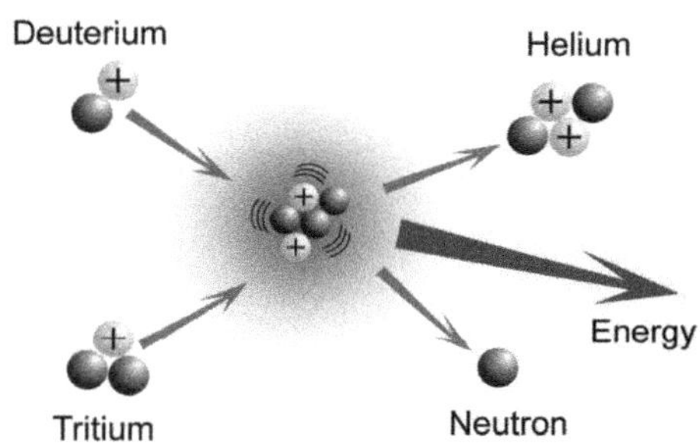

Figure 4.10 Fusion reaction of Li^6.

Lithium has two major isotopes: lithium-6, which comprises 7.4% of the atom, and lithium-7, the remaining 92.6%. When lithium-6 is bombarded with electrons, it reacts to produce helium and tritium plus 4.8 Mev of energy. The deuterium–tritium fusion yields an alpha particle(^{4}He) and a neutron plus 17.6 MeV of energy.

As a mica, lepidolite is very soft (2.5–3 hardness) and has been used in the carving of beads, African animals, and some very fancy sculptures.

4.4 Petalite ($LiAlSi_4O_{10}$)

Is a lithium aluminum phyllosilicate a monoclinic structure. It is the mineral discovered in Sweden (Utoe) in which lithium was originally identified by Arfwedson.

Figure 4.11 Petalite crystal.

Its color is grayish-white and more rarely pinkish. It has two cleavage directions that form an angle of 38.5°. The basal cleavage is perfect.

The theoretical lithium content of petalite is 2.27% Li, but in actual commercial deposits, lithium varies between 1.6% to 2.1% Li. New data confirm that most petalites are slightly Li-deficient, Al-excessive, and (OH)-bearing with respect to the idealized formula $LiAlSi_4O_{10}$ [11]. Its low iron content makes it a desirable mineral for use in pyroceramics.

Figure 4.12 Petalite cut gems.

With a hardness of 6 to 6.5, faceted petalite gems and specimens are prized by gemstone and mineral collectors. However, faceted gems aren't often seen due to petalite's cleavage planes, which makes it difficult to cut. Colorless material is common, and large crystals have been mined. Other petalite colors include white to yellow or gray, yellowish-green to light green and pink. Crystals range from transparent to translucent and are prized for their good clarity.

Petalite occurs in the Karibib area (Namibia), Aracuai (Brazil), Londonderry in Australia, Eastern Transbaikalia (Russia), and Utoe (Sweden). Recently, in Canada, high-quality petalite concentrates have been produced from the Big Whopper pegmatite by Avalon Ventures. The best-known petalite deposit occurs in Bikita (Zimbabwe).

4.5 Amblygonite [$LiAl(PO_4)(F,OH)$]

It is a triclinic mineral and the fluorine-rich end member of the phosphate series, while the rarer montebrasite $LiAlPO_4(OH)$ represents the hydroxyl-rich end member. Tavorite, $LiFe^{3+}(PO_4)(OH)$ with a theoretical lithium concentration of 3.97% (Hudson Institute of Mineralogy), is the iron-rich hydroxyl member. The mineral was first discovered in Saxony, Germany, by August Breithaupt in 1817 and named by him from the Greek word *amblus*, meaning "blunt," and *gonia*, meaning "angle," because of the obtuse angle between the cleavages. Montebrasite was named for the type locality of Montebras, France [12].

Figure 4.13 Amblygonite specimen.

Although amblygonite theoretically contains 4.74% Li, commercial deposits usually carry 3.5% to 4.2% Li. It occurs in lithium- and phosphate-rich granitic pegmatites, together with spodumene, lepidolite, and tourmaline. Its large, white, translucent masses are often mistaken for albite. While amblygonite has been identified in practically every country, it has been commercially mined at Keystone, South Dakota, USA, and also in South Africa, Zimbabwe, and several other countries.

Figure 4.14 Cut amblygonite gem.

It is no longer a commercial source of lithium. Although it is a soft mineral, a clear variety of it from Hebron, Maine, has been faceted as a gemstone.

4.6 Eucryptite ($LiAlSiO_4$)

Eucryptite is lithium alumino-silicate which is deficient in silica. It was named in 1880 by George Brush and Edward Dana from the Greek word for "well concealed," in allusion to its occurrence embedded in albite. It is also known as α-eucryptite or alpha-eucryptite.

It is a mineral related to the beryllium silicate phenakite (Be_2SiO_4) and the zinc silicate willemite (Zn_2SiO_4). It fluoresces under UV light.

While it may contain up to 5.53% Li, the only large deposit is in Bikita, Zimbabwe, along with substantial resources of spodumene and petalite (the lepidolite was exhausted when mined to produce the hydrogen bomb). It was first described in 1880 for an occurrence at its type locality, Branchville, Connecticut [1].

Figure 4.15 Fluorescence in amblygonite.

Eucryptite is a lithium aluminum silicate that is deficient in silica. It has a formula $LiAlSiO_4$ and may contain 5.53% Li. The only large deposit of eucryptite is found in Zimbabwe (Bikita), where its occurrence with quartz suggests spodumene origin [18]. The grade of the eucryptite is 2.34% Li. It has also been reported in Connecticut (Branchville Mica Mine), New Mexico (Harding Mine), Manitoba (Tanco Mine), and Ontario Canada (Nakima Mine), and North Carolina (Foote Mine).

4.7 Hectorite [$Na_{0.3}(Mg,Li)_3(Si_4O_{10})(F,OH)_2$]

Hectorite is a lithium-bearing member of the smectite clay mineral group that includes montmorillonite, beidellite, nontronite, and saponite. The best known pure hectorite deposit occurs in San Bernadino County, California. The name hectorite was proposed for the magnesian bentonite from Hector, California, described by Foshag and Woodford (1936). It is regarded as being the magnesium endmember of the montmorillonite group. No reference was made at that time of its 0.53% lithium value. Hectorite has a trioctahedral structure where magnesium and lithium have replaced aluminum in the octahedral site of the clay structure. This makes it susceptible to swelling [19], a property that has resulted in various formulations. Besides its application in paints and coatings, hectorite clays find use in ceramics laundry, cosmetics, medicine, pottery, porcelain, etc. A further famous application is water treatment due to the strong adsorption power of the platelet surfaces and it is used in face and skin creams, as well as deodorants, cosmetic foundations, and lipsticks.

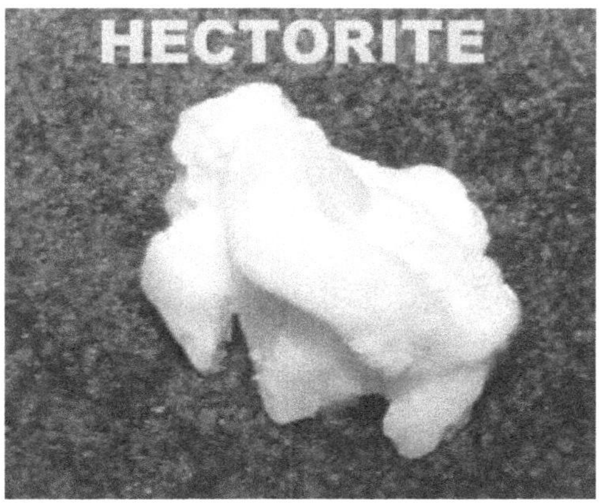

Figure 4.16 Pure hectorite clay.

It has been subsequently identified by the author in Clayton Valley, Nevada, where it occurs in the Tertiary sediments as an alteration product of volcanic ash [13]. As a result of intense exploration for lithium in the many dry lakes of Nevada, more hectorite mineral locations have been identified. Western Lithium Corp. identified hectorite in the McDermitt Caldera in Northern Nevada. Cypress Development intersected additional hectorite clays in a 2018 drilling campaign conducted in the Tertiary Esmeralda formation east of Clayton Valley, Nevada, where the author originally reported its occurrence [14]. Bancora Lithium reported a minor occurrence of hectorite admixed with montmorillonite in a clay prospect in Eastern Sonora, Mexico, averaging 0.35% lithium.

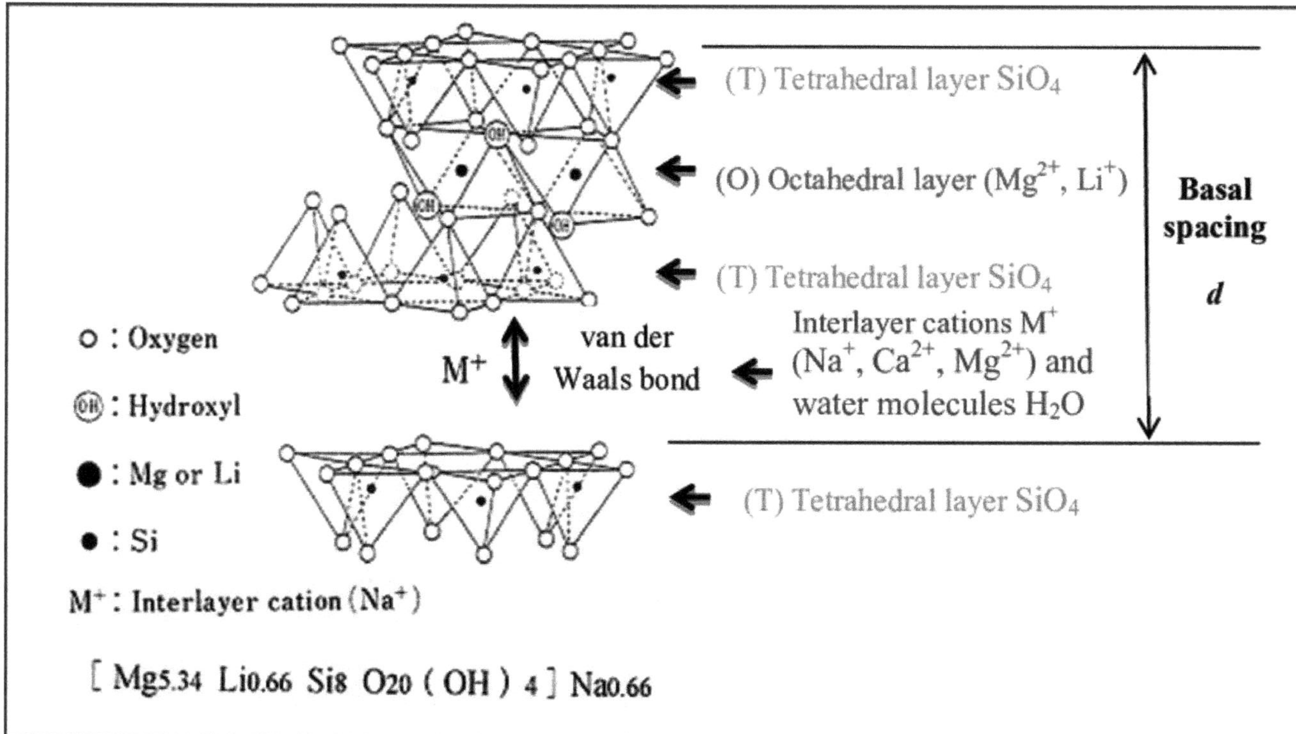

Figure 4.17 Hectorite structure.

4.8 Switzerite [$KLiFe^{2+}Al_2Si_3O_{10}F_{1.5}(OH)_{0.5}$]

It is a trioctahedral mica, an intermediate member of the siderophyllite–polylithionite series, which may contain up to 1.9 wt.% lithium. It was the very first source of lithium mined by Metallgesellschaft of Germany in the Erzgebirge-Fichtelgebirge Anticlinorium, the eastern part of Germany and the northwestern part of Czech Republic, where the Erzgebirge Mountains are called Krušné Hory. The company produced a concentrate of 2.7% Li_2O to recover lithium salts necessary to form a lithium–aluminum alloy called "Scleron" [15].

Figure 4.18 Switzerite crystals.

Originally mined for tin, historical archives indicate that the first adit (stollen) was driven in 1686. The zinnwaldite–albite granite (ZAG) is, on average, composed of plagioclase (albite 34.8%), quartz (32.8%), orthoclase (23.4%), Li-mica zinnwaldite (5.9%), sericite (2.1%) and accessory topaz, fluorite, zircon, cassiterite, and clay minerals.

4.9 Jadarite [$LiNaSiB_3O_7(OH)$]

Recently, a new lithium–boron mineral has been identified near Jadar Valley, Loznica, Mačva District, Central Serbia, Serbia [1]. Rio Tinto exploration geologists discovered the mineral in December 2004 as small, rounded nodules in drill core, and were unable to match it with previously known minerals. The monoclinic mineral was analyzed to contain 7.28% of lithium oxide and 47.12% of boron oxide, a very promising combination. The mineral resources were estimated at 114.5 million tons with an average content of the profitable components averaging 1.8% of lithium oxide and 13.1% of boron oxide [16].

Figure 4.19 Sample of jadarite.

It has been widely reported in the media that this mineral has a similar composition to the mythical "Kryptonite" of Superman fame. So far, the effects of jadarite on superheroes or on exploration geologists for that matter have not been noted by researchers. Jadarite was confirmed as a new mineral after scientists at the Natural History Museum in London and the National Research Council of Canada conducted tests on it [17].

Recently a claystone deposit was identified on the Silver Peak Range, Esmeralda County Nevada, where lithium was identified in a mineral-like searlesite—$NaBSi_2O_5(OH)_2$. The deposit has an average lithium grade of 1800 ppm Li (0.39% Li_2O) and 17,300 ppm boron (5.58% B_2O_3). The mineral is possibly a lower-grade jadarite.

References

1. Mindat.org. The Hudson Institute of Mineralogy.
2. Grew, E. S., Hystad, G., Toapanta, M. P. C., Eleish, A. Ostroverkhova, A., Golden, J., Hazen, R. M., Lithium mineral evolution and ecology:

comparison with boron and beryllium, *European Journal of Mineralogy*, **31**(4), 755–774. doi: https://doi.org/10.1127/ejm/2019/0031-2862

3. Norton, J. J., and Schlegel, D. M., 1955, Lithium resources of North America, *USGS Bulletin* 1027-G.
4. Cerny, P., and Ferguson, R. B., 1972, The Tanco pegmatite at Bernic Lake, Manitoba; IV, Petalite and Spodumene relations, *The Canadian Miineralogist*, **11**(3), 660–678.
5. Heinrich, E. Wm., Salotti, C. A., and Giardini, A. A., 1977, Hydrogen–mineral reactions and their application to the removal of iron from spodumene. *Energy*, **3**, 273–279.
6. Partington, G. A., and McNaughton, N. J., 1995, A review of the geology, mineralization, and geochronology of the Greenbushes pegmatite, western Australia, *Economic Geology*, **90**(3), 616–635.
7. Weeks, M. E., 1960, Discovery of the elements, *Journal of Chemical Education*,VI, **9**, 631–634.
8. Deer, W. A., Howie, R. A., and Zussman, J., 1962, *Rock-forming Minerals*, Vol. 3: Sheet silicates, Longman: London.
9. Sinkankas, J., 1964, *Mineralogy*, Chapman and Hall: New York.
10. Kirchhoff, G., and Bunsen, R., 1861, *Annalen der Physik und Chemie*, **189**(7), 337–338.
11. Cerny, P., and London, D., 1983, Crystal chemistry and stability of petalite, *Tschermaks Mineralogische und Petrographische Mitteilungen*, **31**, 81–96.
12. Foshag, W. F., and Woodford, A. O., 1936, Bentonitic magnesian clay mineral from California. *American Mineralogist*, **21**, 238–244.
13. Anthony, J. W., Bideaux, R. A., Bladh, K. W., and Nichols, M. C. (Eds.), 1995, *Handbook of Mineralogy*, Volume II, Mineralogical Society of America: Chantilly, VA.
14. Kunasz, I. A., 1970, *Geology and Geochemistry of the Lithium Deposit in Clayton Valley, Esmeralda County, Nevada*, Pennsylvania State University thesis.
15. Metallgesellschaft, *Review of the Activities*, No. 6, 1963.
16. Rio Tinto, 2020, *Rio Tinto progresses the Jadar lithium project to feasibility study stage*.
17. Carpenter, G. J. C., and Whitfield, P. S., 2007, Jadarite, $LiNaSiB_3O_7(OH)$, a new mineral species from the Jadar Basin, Serbia, *European Journal of Mineralogy*, **19**(4), 575–580.

18. Westenberger, H., 1963, The lithium minerals, their formation and occrurrences, *Metallgesellschaft Review of the Activities*, **6**, 16–21.

19. Ames, L. L., Sand, L. B., Goldich, S. S., 1958, A contribution on the Hector, California bentonite deposit, *Economic Geology*, **53**(1), 22–37.

Chapter 5

Lithium Applications

From very humble beginnings, lithium is being used in a variety of applications. At the turn of the century, the first application of lithium hydroxide was in Thomas Edison's battery. In 1918, lithium metal was used as an alloy by Metallgesellschaft in the production of Bahnmetal for rail applications. Lithium hydroxide surged with the Atomic Energy Commission lithium hydrogen bomb program, which imported lepidolite from Southern Rhodesia (today Zimbabwe) to extract the lithium 6 isotope. After the cancellation of the program due to the creation of the atom bomb, the lithium industry needed to diversify into new applications to survive. Diversification led to the application of lithium in pyro-ceramics (Corning), lubricating greases (air force), air conditioning, synthetic rubber, and batteries for the electronic industry. These were the main products for several years.

Today, another major development, the electrification of automobiles and energy storage has resulted in a major expansion in lithium production both from pegmatites, brines, and potentially other sources (clays, geothermal). The following figure illustrates the flow of products from raw materials to various major applications of lithium:

- Lithium stearate is mixed with oils to make all-purpose and high-temperature lubricants.

The Lithium Legacy
Ihor Kunasz

ISBN 978-981-4968-31-7 (Hardcover), 978-1-003-37236-3 (eBook)
www.jennystanford.com

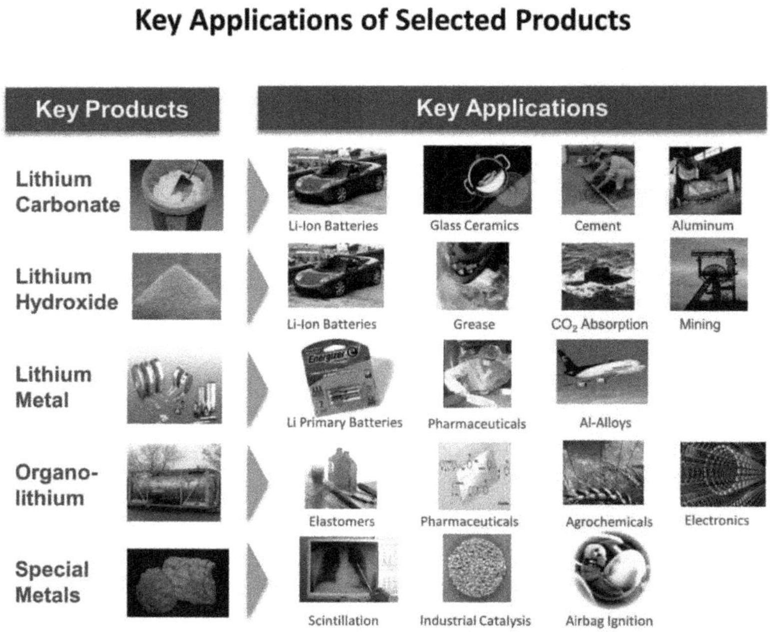

Figure 5.1 Key lithium applications [1].

- Lithium hydroxide is used to absorb carbon dioxide in space vehicles.
- Lithium is alloyed with aluminum, copper, manganese, and cadmium to make high-performance alloys for aircraft.
- Compounds such as $LiAlH_4$ and organolithium reagents (LiMe, LiPh, etc.) are very important as reagents in organic chemistry.
- Lithium metal has the highest specific heat of any solid element and is therefore used in heat transfer applications.
- Lithium is sometimes used as battery anode material (high electrochemical potential) and lithium compounds are used in dry cells and storage batteries.
- Lithium is used in the manufacture of special high-strength glasses and ceramics.
- Sometimes, lithium-based compounds such as lithium carbonate (Li_2CO_3) are used as drugs to treat manic-depressive disorders.

- Specialty inorganic chemicals produced from lithium carbonate are listed below:

Chemical	Formula
Lithium acetate	$LiC_2H_3O_2.2H_2O$
Lithium aluminate	$LiAlO_2$
Lithium aluminum hydride	$LiAlH_4$
Lithium amide	$LiNH_2$
Lithium metaborate	$LiBO_2$
Lithium tetraborate	$LiBO_4$
Lithium bromide	LiBr
Lithium borohydride	$LiBH_4$
Lithium iodide	$LiI.3H_2O$
Lithium nitrate	$LiNO_3$
Lithium nitride	Li_3N
Lithium sulphate	$Li_2SO_4.H_2O$
Lithium silicate	$Li_2O.2SiO_2$
Lithium stearate	$Li_2(C_{17}H_{35}CO_2)$
Lithium phosphate	Li_3PO_4

Besides the small amount of lithium hydroxide used in 1903 by Thomas Edison in his early efforts at batteries, the first large use of lithium was by Metallgesellschaft of Frankfurt/Main, Germany. Following World War, I, Germany had difficulties finding a substitute for the tin lost through heat generation in the axle boxes of railway wagons. A that time, some 7000 tons of tin were required. Various substitutes were evaluated. The "Lurgi" metal consisted of mixtures of barium, calcium, and sodium. A second alloy consisting of 0.7% calcium, 0.6% sodium, and 0.04% lithium, which came to be known as "Bahnmetal" (railway metal) was developed. As the speed of trains increased after 1955, the German Railways changed to roller bearings and abandoned the use of Bahnmetal. Around 1918, it was discovered that lithium had a hardening effect on aluminum. It was thought that this new alloy, named "Scleron," would replace much of the structural steel used in construction and even in the production of airplanes, reducing the total weight by about 10%.

Metallgesellschaft realized that there were not enough lithium raw materials to produce this alloy and decided to start exploiting the necessary raw materials themselves. The ore came from Zinnwald area in SE Germany and the company produced a concentrate containing 2.7% Li_2O, to manufacture lithium salts which were used to produce lithium metal in 1925. The company diversified into lithium carbonate to produce lithium hydroxide and pharmaceutical lithium salts [2].

Although efforts were made to produce a lithium–aluminum alloy to potentially replace steel, in the 1970s, Foote Mineral Company planned to invest some $15 million in such a plant, which never materialized (private communication).

5.1 Lithium Carbonate

After the termination of the Atomic Energy Commission (AEC) contract, one of the more attractive areas was to use increased quantities of lithium (from minerals such as petalite, low-iron spodumene concentrates and lithium carbonate) in pyro-ceramics.

Normally any glass expands upon heating and contracts upon cooling, and if quenched, it disintegrates. Lithium addition counteracts such expansion and contraction; in fact, the addition of lithium results in a 10^{-7} cm/cm negative coefficient of expansion upon heating. The interest in the effect of lithium in glass was initiated at the Pennsylvania State University where Dr. W. Weyl of the College of Mineral Industries directed the research on the behavior of lithium ion in glass. This research was sponsored by the American Lithium Institute, created in 1957 with the participation of Foote Mineral company, American Potash and Chemical Corporation, and Lithium Corporation of America.

In 1962, Dr. G. H. Beal of Corning Inc., New York, related how Don Stookey, the inventor of glass ceramics and John MacDowell had worked on the controlled crystallization of glass. In fact, he was surprised that three glass-ceramics had already been commercialized: lithium-silicate, magnesium aluminosilicate, and lithium–aluminosilicate glass. It is interesting to note that it was a serendipitous experiment that produced the first glass-ceramic. Stookey was experimenting with lithium silicate photosensitive glass containing silver and planning to heat the glass to 450°C. The furnace got overheated to 850°C, and to his surprise, the result was not the expected melted gob but a dense white crystalline ceramic of the same size as the original sample. Silver was an effective nucleating agent for the creation of lithium disilicate ($Li_2Si_2O_5$). The ceramic produced was uniform and nonporous and apparently much stronger than the original glass [3]. Further research showed that viscosity during crystallization is influenced by temperature and phase assemblage, which control the shape of the final product, and that thermal history has a marked influence on the crack resistance of the glass ceramic. This was Corning's success in producing the widely used Corning ware products. George Edwards was the chief geologist for Corning, responsible for locating the best sources of lithium to be used in this application; George was also the principal mover to organize the very first lithium symposium in 1977.

5.2 Lithium Hydroxide

Lithium soaps are produced by saponification of triglycerides, using lithium hydroxide or lithium carbonate as the saponification agent. Lithium soaps are used as lubricant components and form-release agents at relatively high temperatures. The main components of lithium soaps are lithium stearate and lithium 12-hydroxystearate.

Apparently, the origin and application were triggered by the United States Air Force requirement for a lubricating grease that would perform under various temperature conditions. Calcium-based stearates would turn to soap at low temperatures and would not be functional as lubricants. Sodium-based stearates would become liquid at higher-temperature experiences in aircraft and thus would not perform adequately. The solution was lithium-based stearates that function as lubricants over a wide range of temperatures. Today, any serious mechanic will buy "white grease," which is the lithium-based stearate, a very functional grease for any lubricating application. Lithium hydroxide, the main chemical component of the greases, is one of the main components of the sales volumes of lithium companies. With the development of lithium batteries for the automotive industry, lithium hydroxide appears to be the preferred intermediary to produce lithium metal.

Lithium hydroxide is produced by reacting lithium carbonate with a base such as calcium hydroxide. Besides its use in breathing rescue equipment in mines and submarines, its main application is in the production of lubricating greases, which are stearates.

From 1954–1963, the Atomic Energy Commission (AEC) produced the Li^6 isotope for thermonuclear weapons program and produced the first hydrogen bomb. The program was abandoned after the creation of atomic bomb. The lithium six isotope was obtained from lepidolite—a lithium ore—imported from Southern Rhodesia (now Zimbabwe). Lepidolite tonnages increased from 26,909 metric tons in 1954 to 93,545, metric tons in 1957 and then decreased to 64,699 metric tons when AEC program ended. American Lithium Chemicals built a $6.6 million conversion plant in San Antonio, Texas to produce lithium hydroxide from which the Li^6 was separated. At the height of production, the company produced 85–95 metric tons of enriched lithium each year for a total of 422.4 metric tons until the program was abandoned [4].

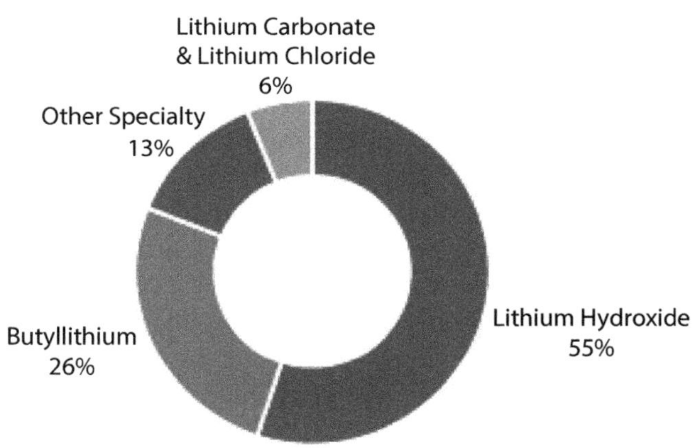

Figure 5.2 Relative lithium applications (%).

After the AEC program, 30,909 metric tons of depleted (Li^7) hydroxide were stored at the Portsmouth, Ohio, facility and 10,455 metric tons of natural lithium hydroxide at the K-25 site. Additionally, 12 metric tons of natural and 8 metric tons of depleted lithium hydroxide were stored at the Y-12 plant, Oak Ridge, Tennessee. The U. S. government offered the depleted lithium for sale. However, the low bids by Foote Mineral Company and Lithium Corporation of America were not accepted and Toxco (the lithium battery recycling company) won the bid.

As illustrated in Fig. 5.2 lithium hydroxide has overtaken lithium carbonate and has become a more important compound in the production of high nickel chemistry [5].

Key advantages of lithium hydroxide battery cathodes compounds include better power density (more battery capacity), longer life cycle, and enhanced safety features.

5.3 Butyl Lithium

The chemical is important in the production of synthetic rubber for automotive tire. Synthetic rubber is any man-made elastomer. Although natural rubber is a renewable resource, it has performance drawbacks: it melts at 180°C and thus cannot be used in high-temperature environments. At low temperature, it becomes brittle and loses its flexibility. Synthetic rubber satisfies these conditions and is thus desirable.

There are many different types of synthetic rubber (neoprene, ethylene propylene diene monomer, styrene butadiene rubber, butyl rubber s-elastomers, silicone, buna N rubber) products for different applications in the automobile sector, including tires, door and window profiles, seals such as O-rings and gaskets, hoses, belts, matting, and flooring. Styrene-butadiene rubber and butadiene rubber (both Buna rubbers) are commonly used for tire manufacture.

When I started my summer fellowship at the Silver Peak Foote Mineral Company brine operation in Clayton Valley, Nevada, the staff, which lived in Tonopah, Nevada, some 50 miles away, would drive every day at a high speed of 90 miles an hour. There was no speed restriction in 1969. At this speed, I remember, new tires were rotated every 3000 miles. What a change since those days! the production of high-quality synthetic rubber now allows today's tire to perform for 60 to 80,000 miles. The secret to this great improvement in performance was the organic component, *t*-butyl lithium. As a catalyst, it resulted in the production of a nearly perfect synthetic rubber of a quality far superior to that which was available to date. Dr. William Novis Smith Jr., Director of research at Foote Mineral Company applied for a patent that described the procedure for making *t*-butyllithium, which was granted to him (US patent 3420903A). His invention relates to an improvement in the preparation of *t*-butyllithium by the reaction of *t*-butyl chloride with finely dispersed lithium (containing a small amount of sodium) whereby the yield of *t*-butyllithium is markedly improved in a controlled and consistent manner.

5.4 Lithium Metal

Lithium is the most important component of batteries. Unfortunately, Li-ion lifespans still aren't particularly long, and experience significant deterioration within their first few years. Five years of extensive use can leave a battery at 70%–90% of the original capacity. Li-ion batteries are still an expensive means of power, with the industry standard hovering around $137 per kilowatt-hour (kWh) in 2020. Tesla's cutting edge NCA battery packs are rumored to be closer to $100/kWh. That being said, costs have come a long way: in 2010, battery prices were $1,100/kWh, representing a 90% drop over ten years. But that decrease is not sustainable over the next decade. Lithium-ion, or Li-ion, is the most prolific battery technology in use today. Li-ion boasts high energy density relative to older nickel-cadmium batteries, and the absence of a memory effect, which causes batteries to lose storage capacity with continued usage. 'Self-discharging'—wherein minuscule chemical reactions in a battery lower its capacity over time—is minimal in Li-ion technology. The solution could be sodium-ion batteries, whose development has recently made astonishing progress. In the foreseeable future, they could replace the lithium-ion batteries currently used not only in electric vehicles, but also in smartphones and laptops.

References

1. https://www.albemarle.com/businesses/lithium/markets--applications
2. Metallgesellschaft, *Review of Activities*, No. 6, 1963.
3. Beall, G. H., 2014, Milestones in glass ceramics: a personal perspective, *International Journal of Applied Glass Science*, **5**(2), 93–103.
4. Schreck, A. E., 1961, Lithium: a materials survey, U.S. Dept. of the Interior, Bureau of Mines circular 8053.
5. Livent Corporation, 2019, Annual report.

Chapter 6

Lithium Pegmatites

Before the discovery of lithium in brines (except for Searles Lake, where lithium was a by-product of processing for borax, potash, soda ash, and salt cake), lithium concentrates and chemicals were produced exclusively from pegmatites. Before the increased interest in lithium production, driven by the expected increase in electrification of automobiles and storage, pegmatites were mined primarily for their tin (cassiterite) and tantalum (columbo–tantalite).

The increased interest in pegmatites resulted in two main research topics:

- The sequence of minerals produced during the crystallization of pegmatites and
- The mechanism of emplacement.

Pegmatites could be simply described as the last "hurrah" of magmatic crystallization.

Depending on the percentage of major components, such as silica, sodium, potassium, calcium, magnesium, and iron, pegmatites can be classified as acidic (granite, rhyolite), intermediate (diorite and andesite), mafic (gabbro, basalt), and ultramafic (peridotite).

Pegmatites are coarse-grained deposits formed within or at the edges of cooling magma. In the case of granites, minor elements such as lithium, beryllium, tin, tungsten, and cesium are not concentrated enough to form their separate minerals in the early stages of cooling. However, as cooling progresses, these unique elements become

The Lithium Legacy
Ihor Kunasz

ISBN 978-981-4968-31-7 (Hardcover), 978-1-003-37236-3 (eBook)
www.jennystanford.com

more concentrated, and as the magma expands and fractures the surrounded rocks, it provides a pathway for exotic minerals such as spodumene, lepidolite, amblygonite, beryl, pollucite, cassiterite, and columbo–tantalite to crystallize.

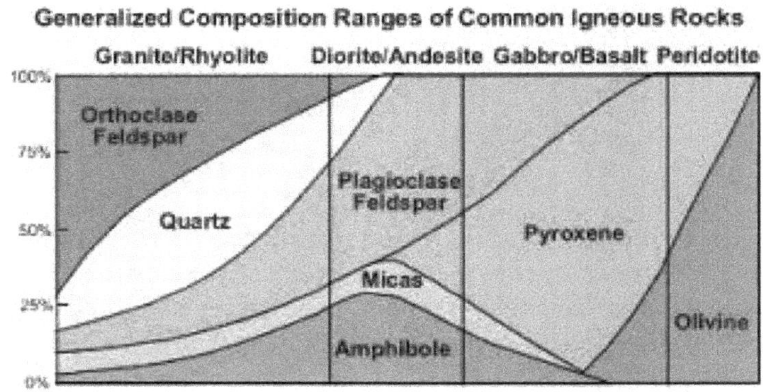

Figure 6.1 Characteristic mineral percentages in various igneous rocks.

A pegmatite is characterized by large crystals mainly mirroring the mineral composition of granites (i.e., quartz, feldspar, and mica). The minor elements (tin, tantalum, tungsten, beryl, cesium) and lithium concentrate in the residual fluids. Jahns pointed to the roles of water and/or other relatively volatile substances, both as a dissolved constituent in granitic magmas and as the dominant constituent of a separate fluid phase that is in the supercritical state under most conditions of pegmatite formation [1].

The field relationship between a granitic parent body and pegmatites has been studied by many investigators (Jahns, R., Cerny, P., London, D.). The earliest documentation on the occurrence of various types of pegmatites were the results of extensive field mapping of the pegmatites in Central Africa and published by Varlamoff in 1958 [2]. Nicholas Varlamoff, a Russian refugee, settled in Belgium in 1923. He obtained a degree in mining engineering at Liege University, followed by a geology degree. From 1934 to 1960, he was a prospector in several African countries (he spoke Swahili in addition to four other languages). In 1960, he became an advisor in the United Nations on Africa and Madagascar while he taught at Queen's College in New York and published 43 articles about rare metal deposits [3]. While he was at the United Nations, I invited Varlamoff to visit Foote's spodumene operation. He offered great insights into the origin of pegmatites, including his interpretation of the Manono–Kitotolo complex.

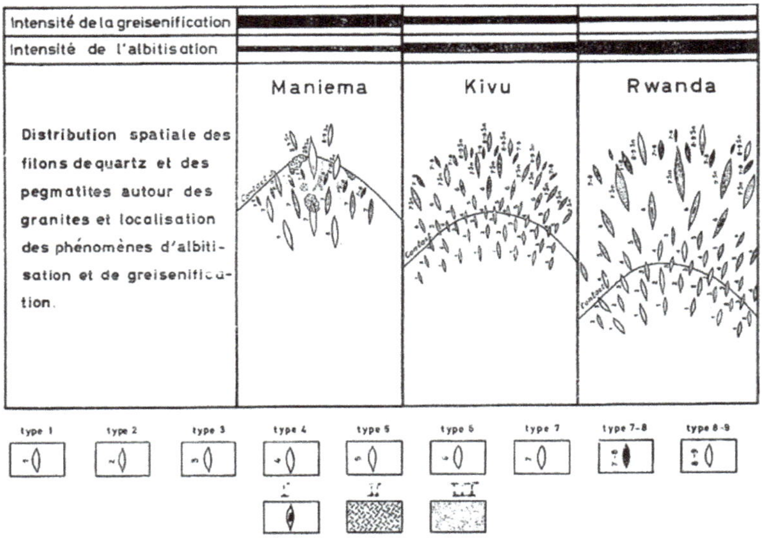

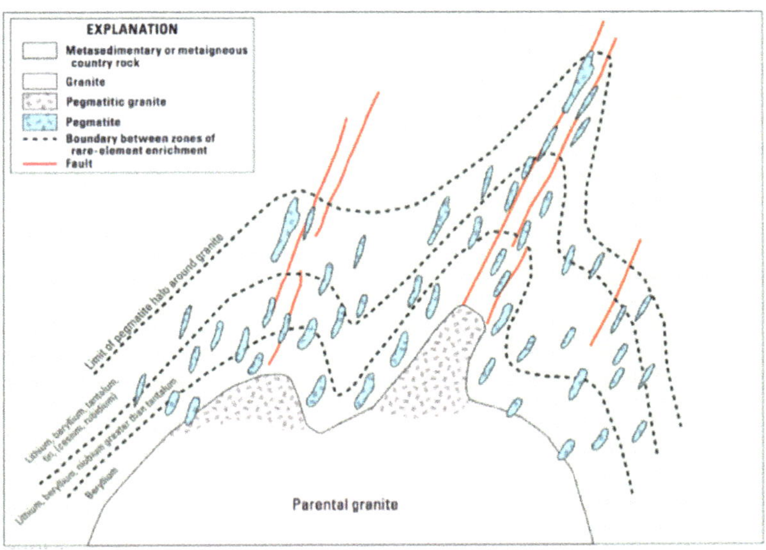

Figure 6.2 Idealized concentric, regional zoning pattern in pegmatite field, adapted from Galeschuk and Vanstone (2005) after Trueman and Cerny (1982). Characteristic rare-element suites of the most enriched pegmatites in each zone are indicated. The most enriched pegmatites tend to occur distally with respect to the parental granite.

The description of the various types of pegmatites has been facilitated by the great topographic relief in that part of the continent. Varlamoff was able to describe the various pegmatites as a function of their distance from the parent magma (Fig. 6.3). This concept was further refined by extensive research and analysis of pegmatites by Cerny illustrating the evolution from barren to beryllium-rich pegmatites to beryllium-niobium-rich pegmatites to complex lithium-cesium-tantalum-tin (LCT) as the distance from the parent magmatic source increases.

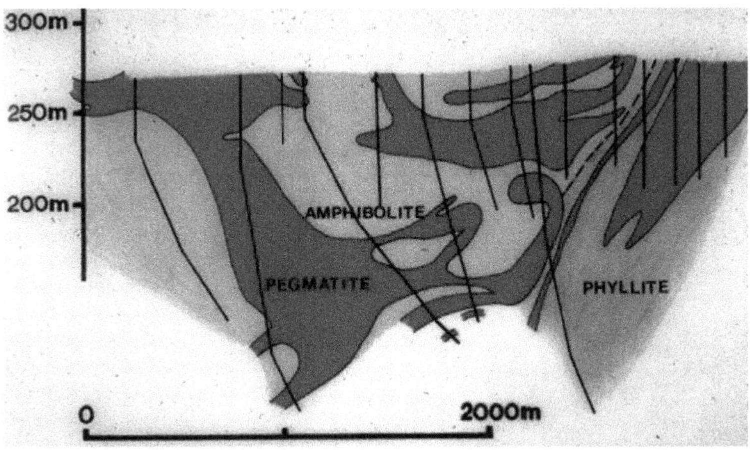

Figure 6.3 Generalized cross-section—Foote Mineral Mine, Kings Mountain North Carolina.

Depending on the stresses either generated by the upwelling magma or regional structural stresses, most pegmatites tend to be linear and subvertical (Canada, Australia). In some cases, as in the case of the pegmatite at the Foote Mineral company at Kings Mountain, North Carolina, they can be highly irregular, reflecting, perhaps, multidirectional weakness zones.

6.1 Unzoned Pegmatites

Unzoned pegmatites make up the bulk of the pegmatite occurrences in the world [4]. There are two basic types of pegmatites: unzoned and zoned.

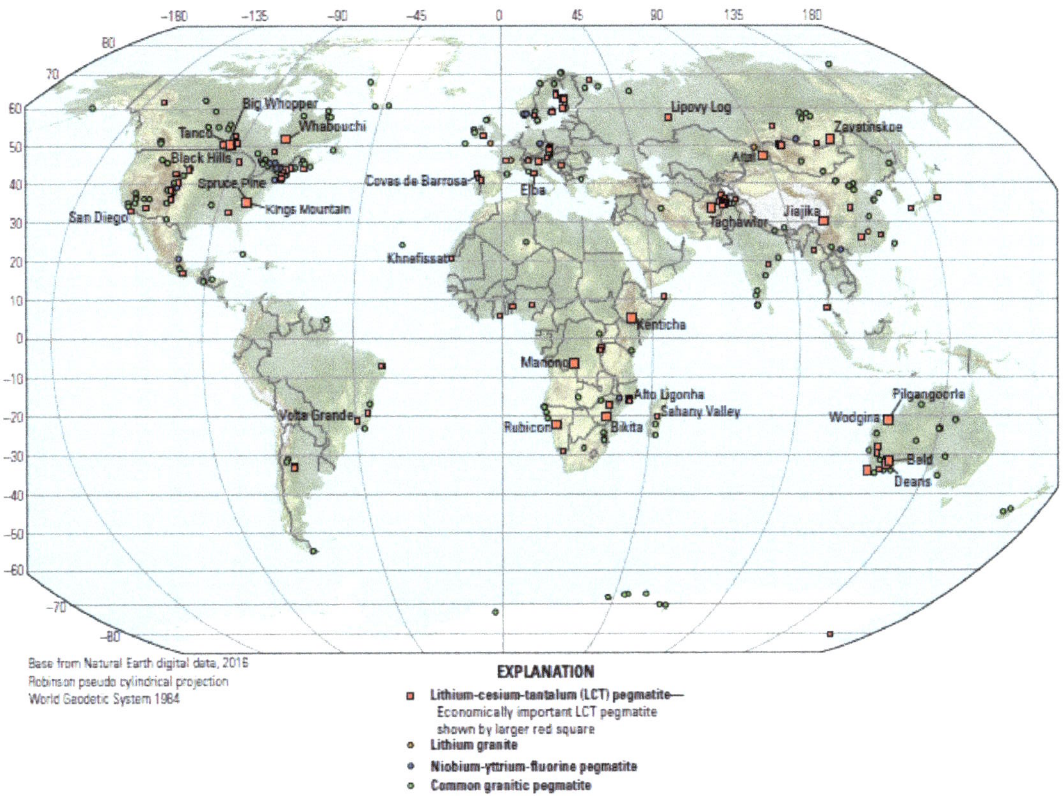

Figure 6.4 Locations of noteworthy pegmatites and related granites categorized by type. Note the widespread distribution of lithium-cesium-tantalum (LCT) pegmatites, which are listed. Economically important LCT pegmatites are shown by larger red squares. Locations for the other types of pegmatites are from Ref. [4].

The simple unzoned pegmatites are exemplified by the Kings Mountain, North Carolina, types. Mining has shown that the pegmatites are injected into greenstones belts and mica schist. The spodumene mineral is fairly evenly distributed in the pegmatite and comprises approximately 20% of the pegmatite body, resulting in an average lithium grade of 0.7% Li. In a number of pegmatites, the lithium grades 0.5% (Mt. Cattlin, Australia, Pervomaisky, Russia, Keliber, Finland). Recent exploration in the tin–spodumene belt has identified additional lithium pegmatites [5]. Spodumene from unzoned pegmatites tends to have a higher iron content (0.6% to 0.9% Fe_2O_3), which is not suitable for ceramic applications, unless the iron can be removed by acid leach after calcination, as was done by Foote Mineral Company. Otherwise, the spodumene concentrates are used as a feed for the chemical conversion into lithium chemicals.

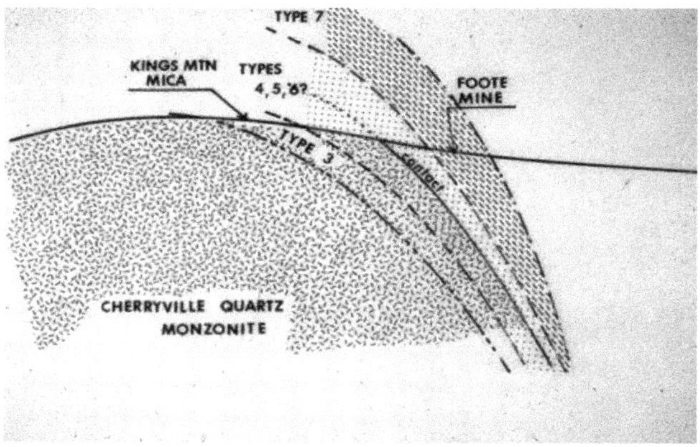

Figure 6.5 Idealized relation between Cherryville Quartz Monzonite Pluton and Associated Pegmatites, Kings Mountain Belt, North Carolina.

The area in which spodumene-bearing pegmatites occur extends from Gaffney, South Carolina, in a northerly direction to Lincolnton, NC, about 16 miles. The zone averages 2 miles in width. Interest in this area was first aroused by the discovery of small amounts of cassiterite in the pegmatites and in greisen, which occurs on the walls of some of the pegmatites. Attempts were made from 1880 into the 1920s to mine and concentrate cassiterite, but there were no successful operations. In 1935, L. M. Williams became interested in the area and began prospecting by trenching and sinking several small shafts. He gradually gained control of a considerable acreage south of the city of Kings Mountain, NC. Mr. Williams states that he produced some ore from a shaft which he shipped to the Maywood Chemical Co. in the late 1930s. In 1937, G. H. Chambers of the Foote Mineral Co. visited the area and examined several properties, but the lithium market was too small to support the necessary concentrating plant [6].

The pegmatites tend to be associated with peraluminous granites, such as the Cherryville quartz monzonite. While most pegmatites tend to be subvertical, the pegmatite body at the Foote Mineral Company operation is highly irregular, pointing to multidirectional stresses that allow for the unique pegmatite geometry.

While pegmatites operations have been suspended in 1997, when Chilean brines supplied lithium carbonate at half the traditional price, there are still large undeveloped resources available for Albemarle and Livent at their existing properties amounting to 59 million tons proven and indicated additional resource of 15 million tons have been estimated to a depth of 500 feet [7]. However, according to the interpretation by Varlamoff, change in composition as a function of the distance from the contact of pluton and at depth may not contain spodumene but might be beryl- and mica-rich [2]. Additional resources based on the recent drilling activities in the tin–spodumene belt may add additional 25 million tons of pegmatites, although the combined grade of 1.09% Li_2O falls short of the historical grade of 1.5% Li_2O mined by the two existing operators, which could restart mining.

6.1.1 Democratic Republic of the Congo (Zaire)

The African continent is richly endowed with rare metal mineralized pegmatites [8]. Central African countries (Democratic Republic of the Congo, Rwanda, Burundi, and Uganda) have an almost 100-year-long history of Sn and Ta production. Further Ta–Nb–Sn provinces are found in Egypt, Ethiopia, Somalia, Mozambique, Madagascar, Namibia, Nigeria, Zimbabwe, and South Africa.

Some of the largest unzoned pegmatites are the Manono and Kitotolo spodumene pegmatites of the Democratic Republic of the Congo (originally Belgian Congo and then Zaire). The pegmatites were discovered in 1910 and exploited for cassiterite. Between 1919 and 1982, a total of 100 million cubic meters of soft pegmatite were processed to produce 185,000 metric tons of cassiterite concentrates.

Because of the constant equatorial rain conditions, the upper 50 meters of the pegmatite has weathered to soft kaolin clay, which made for easy mining by just scrapers and recovery of the tin by heavy media.

In 1978, Phil Comer, vice-president of Foote Mineral Company, and I visited the Manono–Kitotolo pegmatite at the invitation of Geomines director Mr. Claeys. From Kinshasa, we flew commercially to Lubumbashi (the old Elizabethville). From there we hired a small plane to take us to Manono. The pilot must have been a fighter pilot because he could not resist diving down and scaring the rhinoceros and giraffes grazing below us in the savannah. Mr. Claeys received us and was very courteous and informative regarding the deposit. Our return was not as smooth. After returning to Lubumbashi, I had all intents to go to the market to pick up some of the famous malachite specimens. However, arriving at the airport, we found out that we were not listed on the scheduled flight. As Mr. Claeys gave me a letter recognizing me as a member of the *Legion des Leopards*, our driver pointed out the person-in-charge at the airport. When I showed him my letter, he was very gracious about solving the issue but indicated that it required a small fee to rebook us. As I had only a large bill, he said not to worry and that his accountant would give me the change. Upon his return, he informed me that the accountant had no change, to which I promptly indicated that he should keep the change. Back with Phil Comer, I asked how I would report this expenditure. He answered, "saving our lives." The plane was packed with African people who brought all sorts of bundles on board, Phil said that this plane would never take off. I quieted him down by mentioning that the pilots were Belgian and that we should have confidence in them. Once in Kinshasa, showing my famous letter, we avoided the check for illegal diamonds, after which Phil Comer said that he will have three martinis at the hotel to get over our experience. The whole trip was facilitated by the fact that I was born in France and that my French (Belgian) language was a great asset.

Figure 6.6 Large Spodumene Crystals—Manono Pegmatite, DRC.

The report of Barzin, "Managing Director of Geomines," states that the "pegmatites are both 5.5 km long and average 400 m wide" [9]. Through drilling, it has been proved that the pegmatite sills are 125 m deep. A calculation using these dimensions shows that resources are enormous. Barzin reports, however, "that up to 50 meters are kaolinized and altered." Over the years, this layer was exploited for tin and tantalum. Barzin, however, concludes that "exploratory drilling ended in pegmatite," indicating the presence of additional resources. As a result of my field observations and the data from Barzin, I calculated a pegmatite resource of 520,00,000 tons, which was included in the 1978 resource report commissioned by the Energy Research and Development Administration [7].

Recently the Australian company AVZ has become involved in the Manono-Kitotolo sector. The company conducted a mapping of the trenching and drilling that generated an exploration target of 1 to 1.2 billion metric tons, grading between1.25% to 1.5% Li_2O for the entire Manono Project and including between 300 to 400 million metric tons for the Roche Dure Pegmatite alone. Once operational, plans are to send 80,000 metric tons of 6% lithium concentrates per month via the port of Dar es Salaam in Tanzania. A mine life of 50 years is predicted [10].

Figure 6.7 Areal View of the Manono Pegmatite, Democratic Republic of the Congo.

The Manono–Kitotolo pegmatites have been emplaced late during the Kibaran orogeny, probably during the transition from orogenic collapse to extensional tectonics. Different pegmatite veins were injected along the foliation, resulting in deposits that extend over a zone more than 15 km long and 800 m wide and forming the Kitotolo deposit in the southwest and the Manono–Kahungwe deposit in the northeast. The Manono–Kitotolo pegmatites have been linked to the reddish Lukushi and M'Pete leucogranites that have been interpreted as examples of the E-type granites in the Kibara belt. This granite generation has been considered as the parental granite of the rare metal mineralization due to their largely similar age and spatial relation. However, this does not imply that the Lukushi and M'Pete leucogranites are specifically the parental granites of the Manono–Kitotolo pegmatites as pegmatite melts can originate from deeper granitic sources [11].

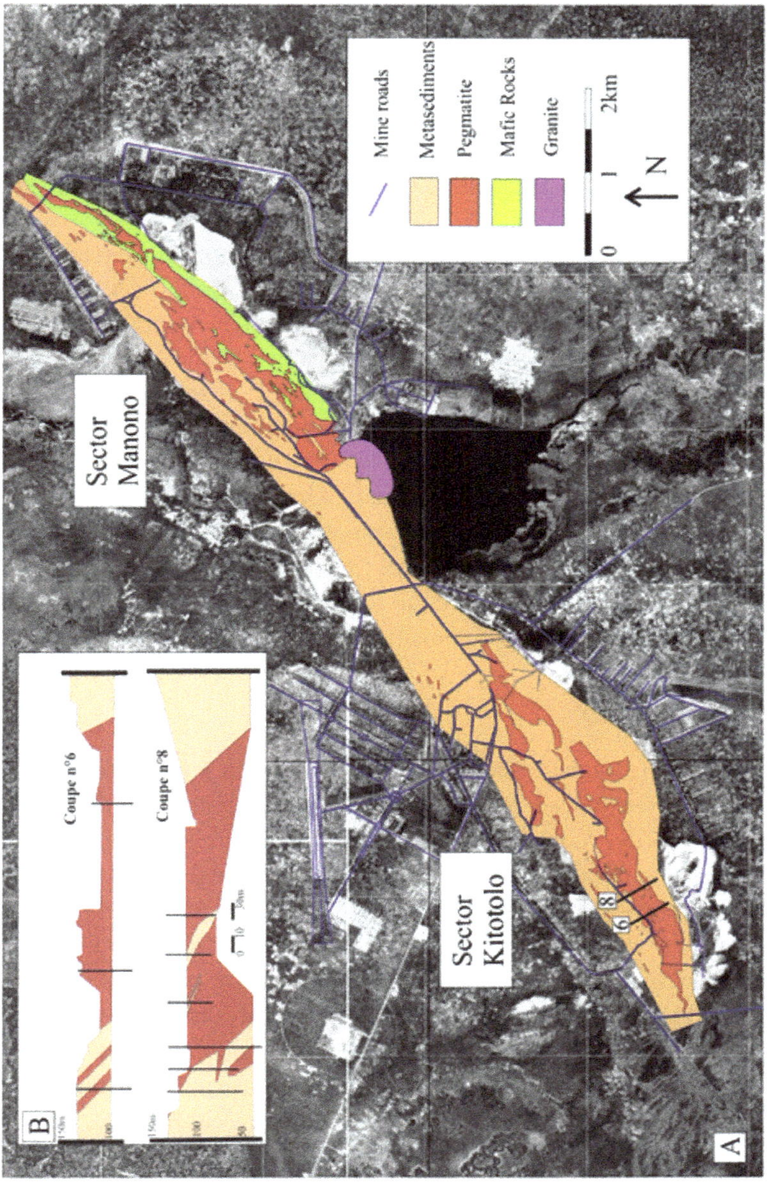

Figure 6.8 Geologic map and cross-sections of the Manono Pegmatite, Democratic Republic of the Congo.

6.1.2 Africa: Mali

The Goulamina, Mali LCT spodumene pegmatite field, located south of Bamako in Mali, is one of the largest hard-rock lithium deposits in the world, with a resource of 32 million metric tons at 1.56% Li_2O [12]. It is interesting to note that Varlamoff, in this review of Central African pegmatites, stated that this area has no economic interest. This was probably because in the 1950s travel was not as efficient as it is today.

Clearly this deposit is a standard lithium pegmatite of average Li_2O concentrations and will likely be used for chemical production. The Chinese have invested in this deposit, explored by Australian lithium miner Firefinch. The announcement states that Ganfeng would acquire offtake rights to 50% of the first-phase annual production capacity of 455,000 metric tons of spodumene concentrate. Kodal Minerals of England has outlined 10 lithium prospects in the Bougouni sector, adjacent to the Goulamina Lithium Project, and has obtained a mining license for a targeted 21.3 million metric tons of measured and indicated resources to produce an annual 220,000 metric tons of a 6% spodumene concentrate, to be purchased by Suay Chin International of Singapore [12].

6.1.3 Europe

While many pegmatites have been identified in Europe, only some have been explored to estimate the lithium resources contained in them. The Weinebene deposits in the Koralpe [13] have been explored in detail. The numerous pegmatites are thin and are likely to be mined by the underground method.

Other spodumene pegmatites have been identified in the Kaustinen region of Finland. Keliber reported plans to produce 140,000 metric tons of spodumene concentrates to produce 15,000 metric tons of lithium hydroxide at the Kokkola plant. The mining reserves (measured open pit and probable underground) are reported at 3.5 million metric tons averaging 1.02% Li_2O. Mining and production have not started.

Cinovec is in the Krušné Hory Mountains, which divide the Czech Republic from the Saxony State of Germany. The resource, a term used for a deposit whose extent is yet to be proven by exploration, could amount to 1.3 million metric tons, or about 3% of the global lithium stock, according to the Czech Geological Survey. One is that ore around Cinovec that contains relatively small amounts of lithium. Jaromir Stary, an expert at the Czech Geological Survey, said it contains four times less lithium in its ore than at the Greenbushes mine in Australia.

A summary of all the European lithium deposits is summarized in the following table, which illustrates that most deposits have too small a resource basis to become important players in the lithium supply chain [14].

Table 6.1 Known European lithium deposits

Description	Mineral	%Li_2O	Reserve (Mt)
EU, Austria, Wolfsberg,	Spodumene	1.0	0.1
EU, Czechia, Cinovec,	Zinnwaldite	0.39	-----
EU, Germany, Sadisdorf	Zinnwaldite	0.45	-----
EU, Finland, Haapaluoma	Spodumene	-----	-----
EU, Finland, Hirvikallio	Spodumene	0.47	0.00047
EU, Finland, Kietyonmaki	Spodumene	0.7	0.007
EU, Finland, Länttä, Ullava	Spodumene	0.94	0.014
EU, Finland, Osterbotten	Spodumene	0.43	0.0019
EU, Finland, Syväjärvi	Spodumene	1.24	1.97
EU, Finland, Rapasaari	Spodumene	1.15	3.46
EU, Finland, Outovesi	Spodumene	1.43	0.28
EU, Finland, Emmes	Spodumene	1.43	0.82
EU, Finland, Leviäkangas	Spodumene	1.01	0.4
EU, Ireland, Leinster	Spodumene	2.3	0.5
EU, Norway, Helgeland	----	-----	-----
EU, Poland, Kostrzya	Zinnwaldite	-----	-----
EU, Portugal, Barroso-Alvao	Spodumene	0.57–1	0.0514
EU, Portugal, Gondiaes	Petalite		

(Continued)

Table 6.1 (*Continued*)

Description	Mineral	%Li_2O	Reserve (Mt)
EU, Portugal, Serra de Arga	-----	-----	-----
EU, Portugal, Barca de Alva	-----	-----	-----
EU, Portugal, Mangualde	-----	-----	-----
EU, Portugal, Guarda-Goncalo	Lepidolite	-----	-----
EU, Portugal, Segura	-----	-----	-----
EU, Serbia, Jadar valley	Jadarite	0.84	1
EU, Spain, Morille	Lepidolite	-----	0.2
EU, Spain, San Jose	Lepidolite	1	0.2
EU, Spain, Alberto	Lepidolite	-----	-----
EU, Sweden, Järkvissle, Medelpad	Spodumene	0.45	0.003
EU, Sweden, Utoe, Haninge	Petalite	-----	-----
EU, Sweden, Varuträsk	Spodumene	1.3–2	0.001
EU, Sweden, Spodumenberget	Spodumene	1.0	-----
EU, UK, Cornwall Camborne	Amblygonite	0.84	-----
EU, UK, Devon	Amblygonite	-----	-----

6.2 Zoned Pegmatites

The character of zoned pegmatites is in the internal geometric arrangement of the lithium-bearing minerals, which is absent in un-zoned pegmatites.

Jahns and Burnham proposed a model based on laboratory and field observations that zoned pegmatites owe their distinctive textural and zonal characteristics to the buoyant separation of aqueous vapor from silicate melt, giving rise to K-rich pegmatitic upper portions and Na-rich aplitic lower zones of individual pegmatites [15].

The zonation and the number of zones are controlled by the intensity of the differentiation of the minerals as a function of their distance from the magmatic parent body.

Zoned pegmatites tend to have low-iron spodumene, usually less than 0.03% Fe_2O_3, which makes them suitable for direct application in pyroceramics. Zoned pegmatites are known in the Black Hills of South Dakota, Bikita, Zimbabwe, Karibib, Namibia, Varustrask, Sweden, San Luis and Sierra Pamapaneas, Argentina, and Bastar and Bihar, India.

The best-zoned pegmatite is represented by the Tanco pegmatite, which is probably the most complex and differentiated known pegmatite [16]. Originally discovered in the 1920s during a diamond drill program, conditions were not right for commercial production until 1969 when Tantalum Mining Corporation of Canada Limited built a 500 ton/day tantalum concentrator. Shipping ceramic grade spodumene concentrate began in 1986. Cabot Corporation acquired 100% of the operation in 1993.

Nine zones have been recognized at the Tanco Rare Element Pegmatite, Southeast Manitoba (Fig. 6.9):

- Border Zone—albite, quartz (tourmaline, apatite)[1]
- Wall Zone—albite, quartz, Li-muscovite, perthite (beryl)
- Aplitic Albite Zone—albite, quartz (Ta oxides, beryl)
- Lower Intermediate Zone—microcline, albite, quartz, spodumene, amblygonite (Li-muscovite, lithiophilite)
- Upper-Intermediate Zone—spodumene, quartz, amblygonite (pollucite, lithiophilite)
- Central Intermediate Zone—microcline, quartz, albite, muscovite (beryl, Ta oxides)
- Quartz Zone—quartz
- Pollucite Zone—pollucite (quartz, spodumene)
- Lepidolite Zone—Li-muscovite, lepidolite, microcline (quartz, beryl)

An additional character of the spodumene in this pegmatite is the spodumene mineral referred to as SQI (spodumene–quartz intergrowth), which has been interpreted as the exsolution of petalite into spodumene and quartz, resulting in a combined lithium oxide concentration of 4.5% similar to that of petalite.

$$\text{Petalite} \Rightarrow \text{spodumene} + \text{quartz}$$

---

[1]Secondary minerals are given in parentheses.

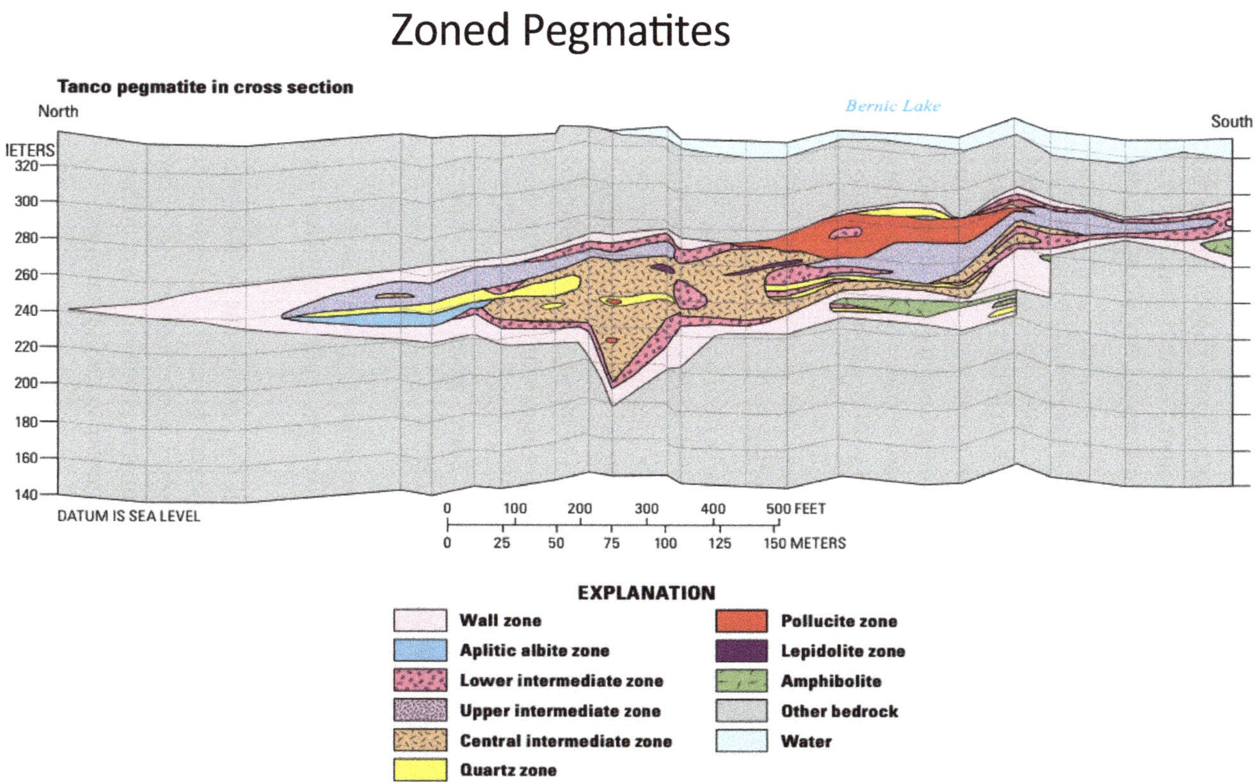

Figure 6.9 Cross section through the Tanco pegmatite, Manitoba, Canada, showing zoning, from stilling and others (2006).

The Tanco pegmatite is one of the major pollucite occurrences.

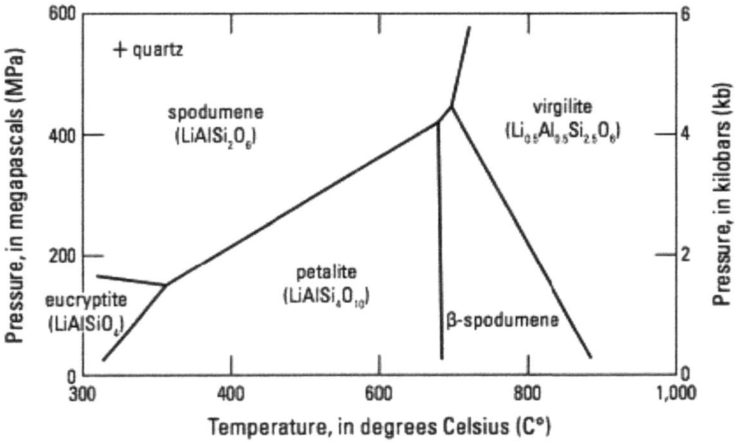

Figure 6.10 Lithium mineral phases as a function of temperature and pressure.

6.2.1 Greenbushes

The largest operational pegmatite is that of Greenbushes in Australia. As has been common for most pegmatites, the pegmatite was originally mined for its tantalum or tin. Then, the geologists realized that the white mineral was not feldspar but spodumene. During my visit to the property, I was shown the cores drilled on the property. Having experience with spodumene core drilling at Kings Mountain, it became clear that some of the intersected white minerals was not feldspar, but that the silkiness exhibited by some of the core intersections could only be spodumene, which I pointed out to the geologist. I would like to think that this was the presumed beginning of the mining of the largest to date spodumene operation.

The Greenbushes pegmatites belong to the lithium-cesium-tantalum family. Believed to be formed approximately 2.5 billion years ago, the Greenbushes deposit hosts pegmatite in a 3 km long and up to 300 m wide main zone. Several small pegmatite dykes and pods are situated along the main body of the pegmatite. The ore body is located within the Balingup Metamorphic Belt (BMB), the southern segment of the Western Gneiss Province. The deposit penetrates the BMB and is hosted within a north to north-west trending lineament called the Donnybrook-Bridgetown Shear Zone. The main mineralization zone features several smaller pegmatite dykes and pods along the main body.

The main Greenbushes pegmatite is 2,500 m long and 61 to 244 m wide, trending north-northwest. It is classed as a large LCT complex pegmatite (spodumene sub-class) and is severely deformed, recrystallized, deeply weathered, and fine-grained to a point where it is difficult to map out the zonation. Two main zones are present:

- Albite/Ta–Sn zone
- Spodumene zone (hanging walls of the main pegmatite)

The lithium ore zones comprise mainly spodumene, apatite, and quartz, with some ore zones returning upward of 5% Li_2O [17].

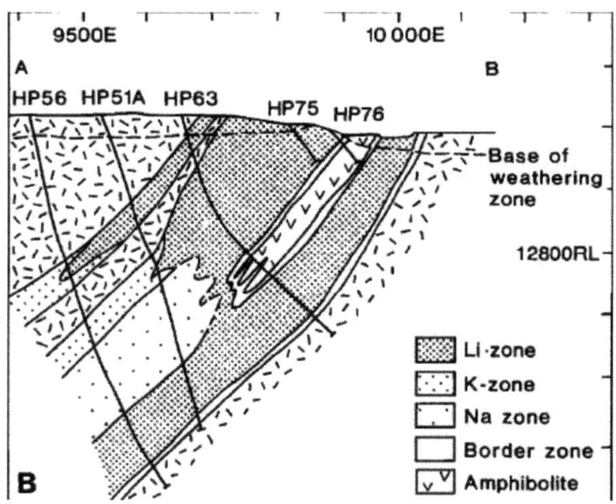

Figure 6.11 Cross section—Greenbushes Spodumene Pegmatite, Australia.

6.2.2 Bikita Minerals

The Bikita granitic pegmatite is located about 65 km east-northeast of Fort Victoria in the Victoria schist belt, surrounded by plutons of Archean granitoid rocks of the Zimbabwean craton. The age of the pegmatite was given as 2650 Ma [19]. The pegmatite body strikes north–northeast and dips moderately to the east, and was emplaced in quartz-bearing amphibolite; it has an exposed length of ~1700 m and a true width variable from 40 to 70 m. The internal zoning of the Bikita pegmatite is quite complex. Cooper (1964) distinguished 9 to 13 zones in separate segments of the pegmatite; he subdivided the tabular body into the border, upper wall, upper-intermediate, core, lower intermediate, and lower wall zones.

Norton (1983) fitted the internal structure at Bikita into his universal scheme of zoning in complex lithium-rich pegmatites. The zoning also varies laterally, with pronounced differences among the intermediate zones between the northern Al Hayat and southern Bikita sectors (the latter, in turn, subdivided into a central quarry and southern-southeastern area sectors) of the pegmatite [20, 21].

Zoning in the Bikita Sector of the Bikita Granitic Pegmatite*

Hanging-wall greens/one Sample numbers
Border zone Selvedge (plagioclase, quartz)
Wall zones Muscovite band WI, W2 Hanging-wall feldspar zone
Intermediate Petalite and feldspar zones
Spodumene (i) massive
(upper) Spodumene (ii) mixed I, 3, 4
Pollucite
Feldspar-<quartz
"All-mix" zone 2, 7, W9, Wl3
Core zones Massive lepidolite[1] (i) high-grade core W7, W8 Massive lepidolite (ii) nearly pure lepidolite Lepidolite-<quartz shell (i) lepidolite W3, W4 Lepidolite-<quartz shell (ii) amblygonite
Intermediate "Cobble" zone
zones Feldspathic lepidolite (lower)
Wall zone Footwall feldspar zone rhythmically

References

1. Jahns, R. H., and Burnham, C. W. C., 1969, Experimental studies of pegmatite genesis. *Economic Geology*, **64**(8), 843–864.
2. Varlamoff, N., 1958, *Les gisements de tungstène au Congo Belge et au Ruanda-Urundi*, Academie Royale des Sciences Coloniales, Tome VIII, Fasc. 2.

3. Tchoumatchenco, *et al.*, 2016, Geologists of Russian origin in the francophone countries, *Boletin Geologico y Minero*, **127**(2/3), 711–738.
4. Bradley, D. C., McCauley, A. D., and Stillings, L. M., 2014, Mineral deposit model for lithium-cesium-tantalum pegmatites, U.S. G. S. Scientific Investigations Report 21010-0507-0.
5. www.piedmontlithium.com
6. Goter, E. R., Hudspeth, W. R., and Rainey, D. L., 1953, Mining and milling of lithium pegmatite at Kings Mountain, NC, *AIME*.
7. Evans, K., 1978, Lithium reserves and resources, *Energy*, **3**(3), 379–385.
8. Von Knorring, O., and Condliffe, E. 1987, Mineralized pegmatites in Africa, *Geological Journal*, **22**, 253–270. https://doi.org/10.1002/gj.3350220619
9. Barzin, H., 1952, Geomines: a major open-pit tin producer in the Belgian Congo, *Engineering and Mining Journal*, **153**(11), 86–89.
10. avzminerals.com.au.
11. Dewaele, S., Hulsbosch, N., Cryns, Y., Boyce, A., Burgess, R., and Muchez, Ph., 2015, Geological setting and timing of the world-class Sn, Nb-Ta, and Li mineralization of Manono–Kitotolo (Katanga, Democratic Republic of Congo). *Ore Geology Reviews*,72, 373–390
12. NS Energybusiness
13. Goed, R., 1989, The spodumene deposit at Weineben/Koralpe/Austria, *Mineralium Deposita*, **24**, 270–278.
14. Global lithium sources, industrial use and future in the electric vehicle industry: a review. http://www.criticalmetals.eu/reports/120964-170531-xCritical-Metals-presentation
15. Jahns, R. H., and Burnham, C. W., 1969, Experimental studies of pegmatite genesis: I. A model for the derivation and crystallization of granitic pegmatites. *Economic Geology*, **64**, 843–864.
16. Cerny, P., Ercit, T. S., Vanstone, T., *et al.*, 1996, *Petrology and Mineralization of the Tanco Rare Element Pegmatite, Southeast Manitoba (FieldTrip A3)*, Geological Association of Canada.
17. Partington, G. A., MsNaughton, N. J., and Williams, I. S., 1995, A review of the geology, mineralization, and geochronology of the Greenbushes pegmatite, Western Australia. *Economic Geology*, **3**, 616–635.
18. Tyndale-Biscoe, R., 1951, The geology of the Bikita tin-field, Southern Rhodesia, *Transactions of the Geological Society of South Africa*, **54**, 11–26.

19. Herzog, L. F., 1960, Age determination by X-ray fluorescence rubidium-strontium ratio measurement in lepidolite, *Science*, **132**(3422), 293–295.
20. Cooper, D. G., 1964, The geology of the Bikita pegmatite, in Haughton, S. H. (ed.), *The Geology of Some Ore Deposits in Southern Africa*, Vol. 2, Geological Society of South Africa: Johannesburg, p. 441–461.
21. Symons, R., 1963, Operation at Bikita Minerals (Pvt.) Ltd., Southern Rhodesia, in *Pegmatites in Southern Rhodesia*, a Symposium, Institute of Mining and Metallurgy: Salisbury, p. 58–72.

Chapter 7

Lithium Brines: Geology and Geochemistry

Brines have been sources of salt for ages. Even early man must have located pools or realized their formation along the sea shores where the glistening crystals attracted his attention and he found the salt not only tasty but also necessary for his well-being. For centuries, seaside saline brines have been harvested for their sodium chloride characteristic flavors and sold as (rather expensive compared to mined salt) food additives in California (San Francisco Bay), France (Ile de Re), Hawaii, Namibia (Walvis Bay), Thailand (Pak Tale, Petchaburi), and India, (Tamil Nadu). Black salts varieties from Cyprus, Hawaii, and India are not naturally occurring but are mixed with fine charcoal to darken the salt. A 1938 English stamp shows the raking of salt in Turks and Caicos Islands.

Various cultures have spent an enormous amount of time reconnoitering areas where salt could be recovered, even in the far reaches of the Saharan Desert where camel trains would take weeks to reach the deposits, workers to dig up slabs of salt, pack them onto camels, and take weeks to bring the "white gold" back to markets. It was only in the late 20th century that brines were treated to recover, in addition to salt, other useful elements such as potassium, magnesium, sulfate, and lithium.

The Lithium Legacy
Ihor Kunasz

ISBN 978-981-4968-31-7 (Hardcover), 978-1-003-37236-3 (eBook)
www.jennystanford.com

Figure 7.1 Stamp illustrating Turks and Caicos Islands hand-salt harvesting.

Economic brines are saline liquids formed because of evaporation of various sources, which impart the characteristics of the final concentrated solutions. However, brines (either of marine or continental origin) produce different sequences of salts.

Marine types, which resulted from the evaporation of ancient seas have produced extensive deposits of sodium chloride (halite) and potassium chloride (sylvinite). Thick deposits such as the Silurian Salina halite deposits underlie Michigan, Ontario, Ohio, and New York as well as the Permian Basin from Texas to Utah. All these resulted from the evaporative concentration of sea water, which consists predominantly of sodium chloride and subordinate amounts of sulfate, magnesium, and potassium and practically no lithium. Seawater contains very small amount of lithium, about 0.017 ppm, and cannot produce a lithium-rich brine. The six most abundant ions of seawater are chloride (Cl^-), sodium (Na^+), sulfate ($SO^2{}_4{}^-$), magnesium (Mg^{2+}), calcium (Ca^{2+}), and potassium (K^+). By weight, these ions make up about 99% of all sea salts. Inorganic carbon, bromide, boron, strontium, and fluoride constitute the other minor dissolved substances of seawater [1]. Evaporation must exceed 90% to reach the point of sodium chloride saturation.

Continental brines, associated with ancient volcanic centers, which acted as enriched sources of lithium, potassium, and boron, have been the centers of lithium production.

The best example of a modern brine that originated from seawater is the Dead Sea. Although it has a lithium concentration of only 20 ppm, it contains significant amounts of potassium, magnesium, and sulfate. The brine is pumped to the Southern Basin where a complex of solar evaporation ponds concentrates the brine. Solar evaporation of this complex seawater brine system precipitates a mixture of carnallite ($KCl.MgCl_2.6H_2O$) and sodium chloride salts. Hot water leaching removes the magnesium and produces high-purity potassium chloride.

Figure 7.2 Satellite view of Dead Sea Potash Pond operations.

On the western side of the Dead Sea, the Israeli Potash Corporation produces a greater variety of the elements present in the Dead Sea brines.

In addition to potassium chloride (11% of world production), the company also produces bromine (which account for 35% of the world production) and 9% of the world magnesium [2].

Brines, such as those from the Great Salt Lake of Utah are residues from large diluted continental seas, which have concentrated over millennia and have reached a lithium concentration of about 70 ppm. Evaporation and shrinking of ancient Lake Bonneville resulted in the present Great Salt Lake and the adjacent Bonneville salt flats (famous for its flatness and use for establishing ever-increasing world speed records).

Approximately 15,000 years ago, lakes in the Great Basin of the western United States and South America achieved their maximum late Pleistocene extents. Three of those lake systems stand out in terms of their enormous dimensions:

- Lake Lahontan, Nevada, with a surface area of 22,300 km^2
- Lake Bonneville, Utah, with a surface area of 51,300 km^2
- Lake Minchin, Bolivia, with a surface area of 60,000 km^2

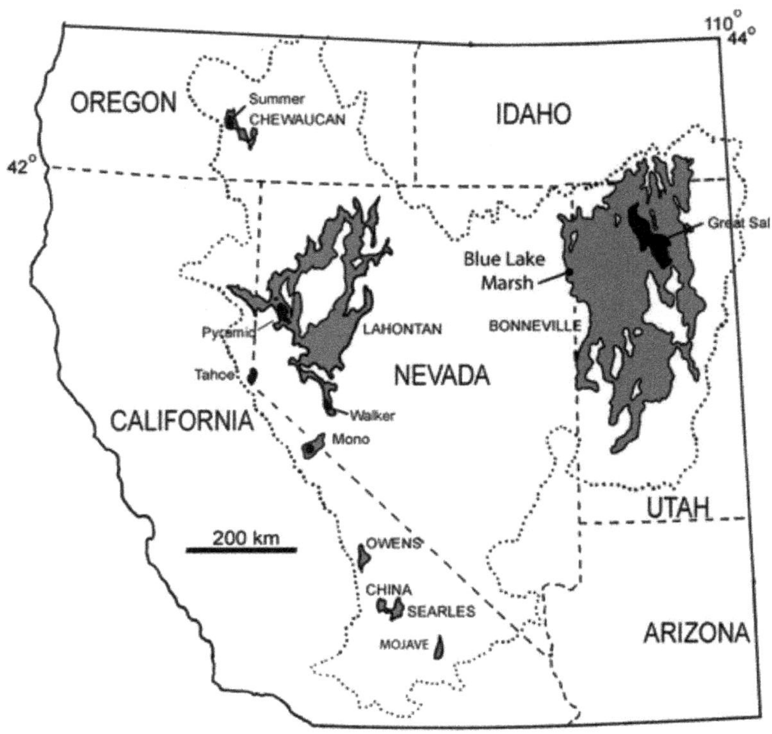

Figure 7.3 Western US Pleistocene Pluvial Lake extensions.

These gigantic freshwater bodies had surface areas approximately 10 times their reconstructed mean historical values. The levels of Lake Bonneville rose and fell several times between 45,000 and 10,500 years ago. [3].

The northern arm of the Great Salt Lake contains 64 ppm Li while the shallower southern arm has 22 ppm Li and the deep southern arm has 43 ppm Li. My thesis advisor on the Clayton Valley lithium system had explained that the lithium concentration in the present Great Salt Lake is the result of the volume reduction from the extensive ancestral Lake Bonneville (1.5×10^{13} m^3) to the volume of the present Great Salt Lake (1.89×10^{10} m^3). One can therefore infer that Lake Bonneville waters probably contained about 0.1 ppm Li.

In May 1967, Great Salt Lake Minerals and Chemicals Corporation (GSL), a joint venture between Lithium Corporation of America (51%) and Chemsalt, the U.S. subsidiary of Salzdetfurth A.G. of Germany, began building a large plant for commercial extraction of potassium sulfate, sodium sulfate, and magnesium chloride, along with common salt. The plant included 17,000 acres of evaporation ponds just north of Little Mountain, west of Ogden on the lake's eastern shore [4]. GSL was to begin production of potassium sulfate and sodium sulfate in late 1970, and production of lithium and magnesium chloride by late 1971 [5]. Unfortunately, the efforts to produce lithium carbonate from the low lithium concentration failed.

There are renewed efforts to recover lithium from the Great Salt Lake brines. On April 25, 2019, US Magnesium LLC and Wollam Construction broke ground on the new lithium carbonate production plant in Rowley, Utah. All major equipment has been installed and initial production is presumably slated for the second quarter of 2020 [6].

Other significant residual lakes and salars are the Salar de Uyuni and Coipasa, remnants of the much larger body of water, Lake Minchin, which existed around 10,000 Before the Present (BP) and covered an area of 120,000 km^2 [7]. Subsequent evaporation shrank the sufaces of the freshwater lake Poopo in the north and the saline brines in the salars of Coipasa and Uyuni and Empexa in the south. While the Great Salt Lake system has a low lithium concentration, volcanic activity on the Andes contributed lithium-rich hydrothermal water to the basins resulting in substantially elevated lithium concentrations as the southern system shrank in size. Lithium concentration in the Salar de Coipasa range from 170 to 580 ppm Li [8]. Lithium in the Salar de Uyuni, which has been extensively sampled, ranges from 80 ppm Li at the edges of the salar to 500 ppm Li in the main body. Lithium values exceeding 2000 ppm occur in the southern portion where the Rio Grande Delta discharges [8]. This sector is the focus of Bolivia's attempt at producing lithium carbonate.

In the lithium chapter of the 1975 edition of *Industrial Minerals and Rocks*, the author called the 1960s decade the "brine decade" [9]. Since then, this contention has held up and has been well supported by the identification of additional lithium-bearing brine deposits in South America and Asia from which lithium and other chemicals are produced [16]. Considering the territorial magnitude of the arid sectors of the South American and Asian continents, it is not surprising that many chemical-rich saline deposits have been identified. Indeed, brines have become the predominant source of lithium carbonate in the world today. (In addition, lithium-rich brines have been identified in association with numerous oil fields and occur in geothermal brines; however, these occurrences are not treated in this book, because their complex chemistry and environments have precluded their economic development). Much research has been done on the chemistry of these brines indicating that although playas and salars are similar in many respects, they nevertheless exhibit individual characteristics. While many saline basins have been identified, three salars have been selected for more detailed discussion, as they represent the three common salars: Clayton Valley, Nevada, USA; the Salar de Atacama, Chile; and the Salar de Uyuni, Bolivia.

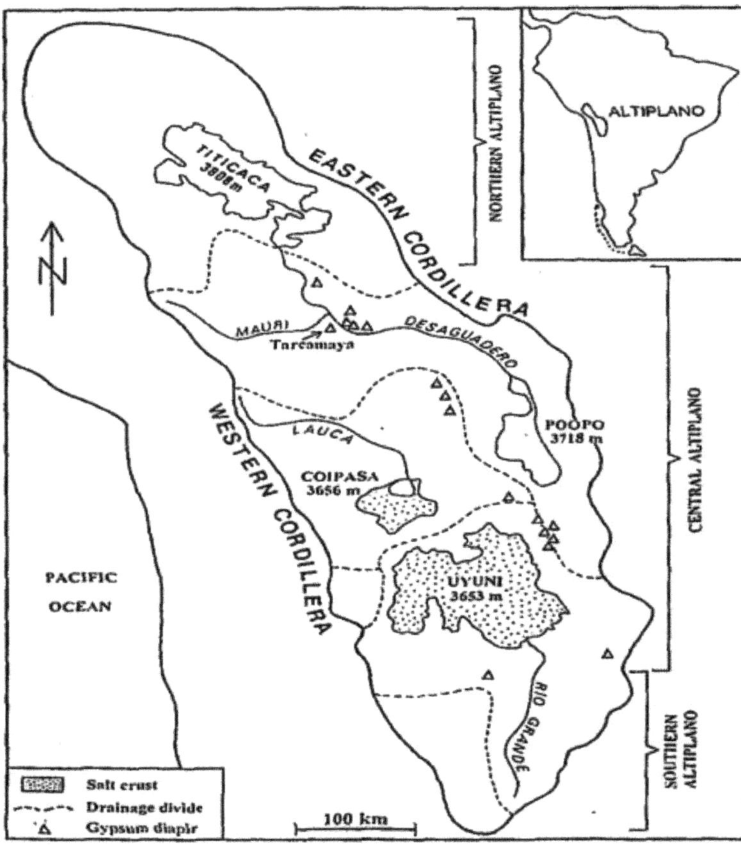

Figure 7.4 Schematic map of the Altiplano showing major drainage basins and gypsum diapirs.

Lithium brines are a particular subset of saline waters that occur in a particular enrichment environment. What are the requirements for a particular lithium brine deposit?

Lithium-bearing salars or desert basins have the following similar characteristics:

- They occur within Tertiary or recent volcanic belts: earlier lithium-rich sediments.
- They are located in closed structural depressions to receive lithium-bearing solutions.
- They occur within the desert areas of the earth, concentrated by natural solar evaporation.

These are the fundamental requirements for the occurrence of economic lithium brine deposits.

For these reasons, lithium brine deposits are in volcanic belts that provide the lithium, in desert regions that facilitate evaporation and concentration, and in structurally unstable regions that provide the depression to contain the discharged geothermal solution loaded with lithium potassium and other elements.

In Fig. 7.5, one will notice that not all deserts have lithium brine deposits. In South Africa and Australia, the desert regions are devoid of recent volcanic activities and therefore lack the hydrothermal lithium source, but, as in the case of Australia, they have significant lithium pegmatite deposits.

One of the most comprehensive review on lithium occurrences and processes was published by Donald Garrett [10] .

The first requirement simply establishes the source of lithium. Studies by the author and colleagues [11, 12] and by many researchers [13] have concluded that the volcanic environment has been the dominant supplier of the lithium either directly through hot springs or geothermal solutions; indirectly through the leaching of lithium-bearing volcanic or clastic sediments; or by the recycling of trapped lithium-bearing solutions.

This condition is met by all three salars discussed. Figure 7.6 illustrates what was known geologically in the 1970s when scant geological information (except for Silver Peak) was available.

- Silver Peak, Nevada
- Salar de Atacama
- Salar de Uyuni

Figure 7.5 Desert belts of the Earth and location of major lithium brine deposits.

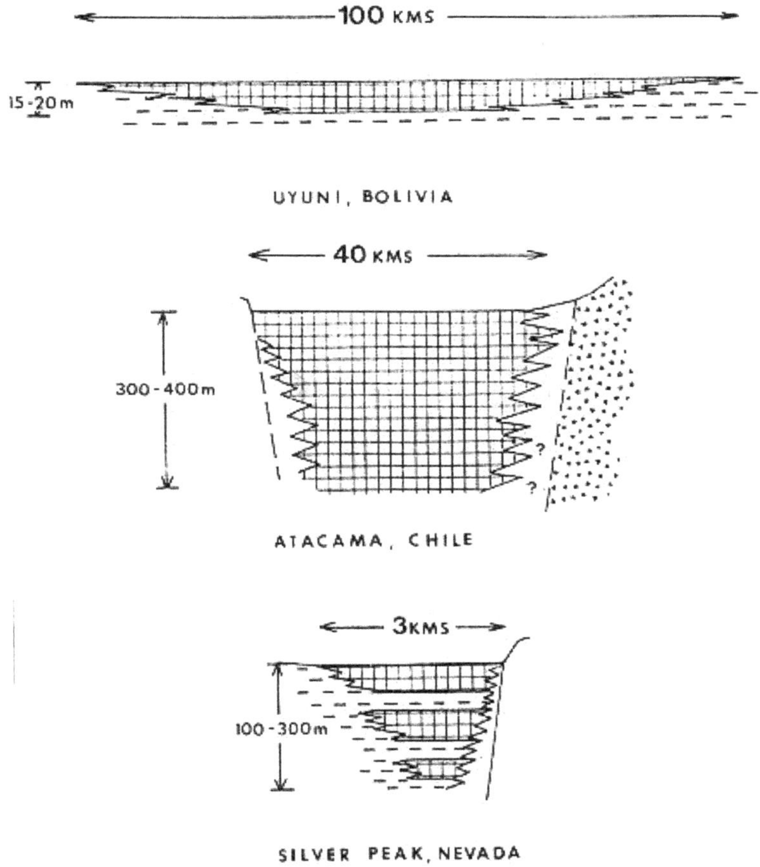

Figure 7.6 Schematic types of basins and depressions.

They all occur in areas with abundant volcanic rocks. Strong structural control is evident in Clayton Valley, Nevada, and in the Salar de Atacama area. Direct structural control is not obvious for Salar de Uyuni. Finally, all potentially "commercial" lithium concentrations are the result of concentration by solar evaporation. Except for the Imperial Valley geothermal field and oil-field brines where lithium concentrations as high as 280 ppm are a result of secondary phenomena, that is, concentration under proper climatological factors. The idealized stratigraphic columns of the three basins indicate significant differences between them as well and reveal their historical evolution. It must be pointed out that the author's interpretation was made in the 1970s when very minimal geological information was available. The general characteristics are nevertheless still valid. Data collected later by drilling deep holes have led to a substantial re-interpretation of the geological nature of these basins.

The hydrothermal solutions accumulate either in asymmetric basins (Silver Peak, Nevada; Salar de Atacama; Salar de Cauchari) or in relatively symmetrical depressions (Uyuni). As the systems evolve, the concentrated elements accumulate by gravity in the deeper portions of the basins, while their concentration decreases toward the edges of the basins where groundwater enters the basins. If the brine is removed by pumping, over time, the saturated brine will be replaced over time by dilute brines. The rate of dilution depends on the rate of brine withdrawal. Houston [14] illustrated the two major types of salars in a graphical summary (Fig. 7.7):

- mature
- immature

In mature salars, the hydrothermal inputs were substantial and the concentration mechanism high enough, such as in the Salar de Atacama and the Salar del Hombre Muerto, to result in a mostly sodium chloride deposit. These salars reflect a high evaporation sequence and resulting in a concentrated lithium brine.

The models presented by Houston are more complex as deep drill-hole information has become available. For example, at the Salar de Uyuni, climatic records indicate several cycles of dryness and wetness with substantial salt deposits from 140,000 to 160,000 and 120,000 to 100,000 years ago and the most recent salt layer that has been used to estimate the earliest lithium resources at the Salar de Uyuni.

Information on original lithium resources in 1980 is based on the extensive work of Risacher of ORSTOM [15], using a salt thickness of 20 meters, which represented the latest uppermost salt layer. However, with the available deep hole data, the resources have been upgraded, although porosities and permeabilities have not been determined to justify the higher lithium resource. It is not clear whether the lower salt, while indicating high evaporation, contains economic concentrations of lithium. However, chemical analyses of the deeper brine indicated that lithium was already an important component of the early system as documented in the following cross-section. Even more important is the fact that lithium-rich brines had already been present 140,000 years ago, indicating it is likely that early weaker solutions were already present and concentrated and the process resulted in fluctuation in lithium strength during the pluvial and interpluvial periods. This also resulted in the largest observable lithium resources evaluated today for their ever-increasing lithium concentrations, leading the concentrations recorded in the latest upper 20 m of the salt.

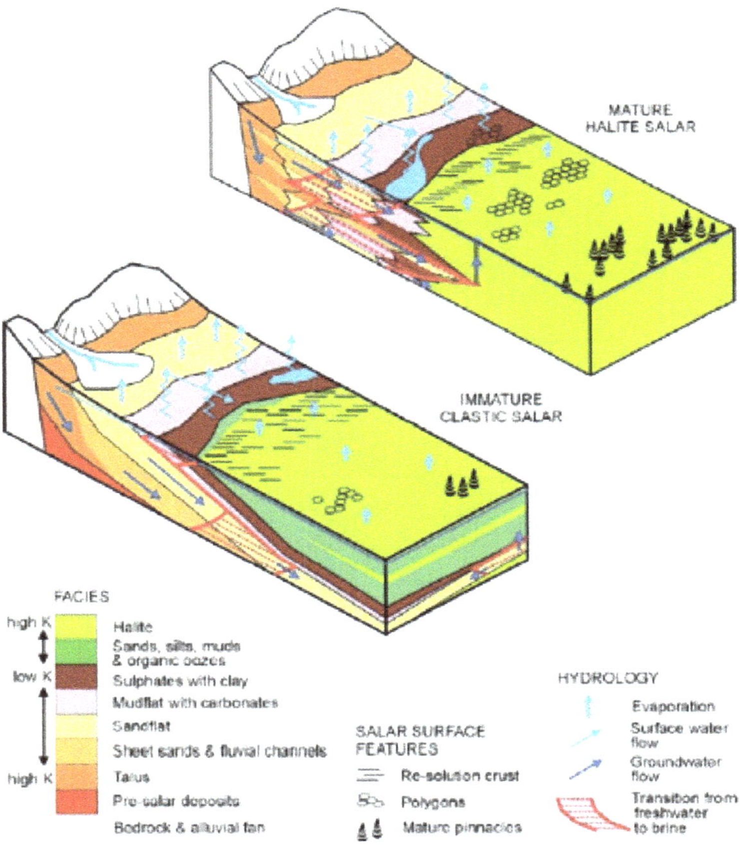

Figure 7.7 Illustration of mature and immature salars (after Houston).

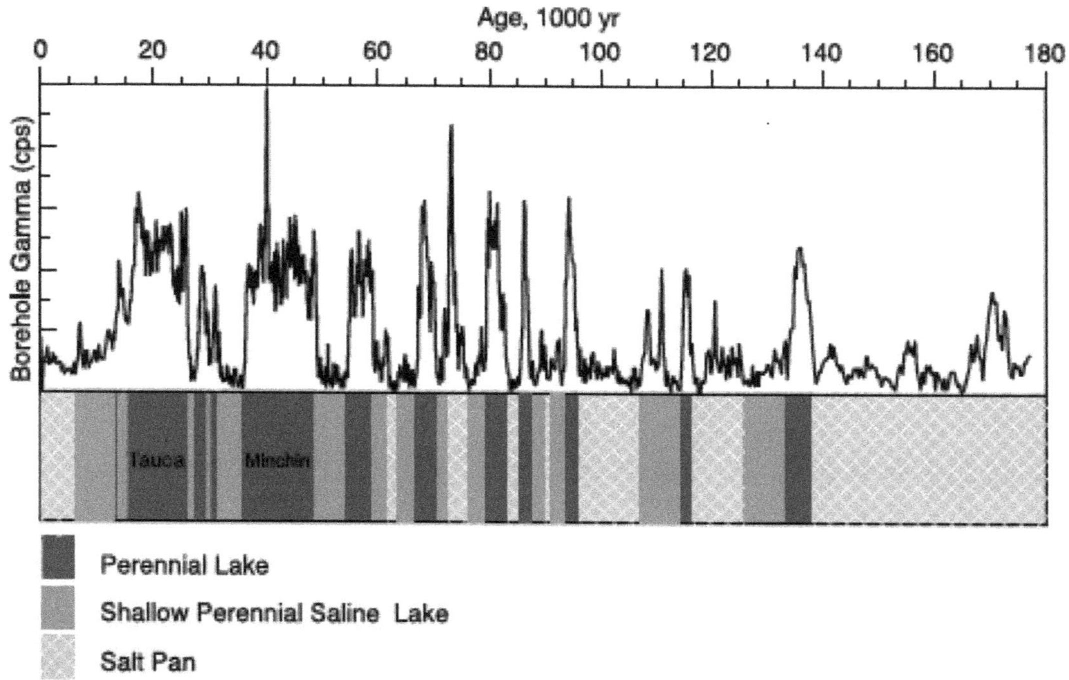

Figure 7.8 Generalized stratigraphy of the upper 100 m of the Salar de Uyuni core displayed versus calendar age.

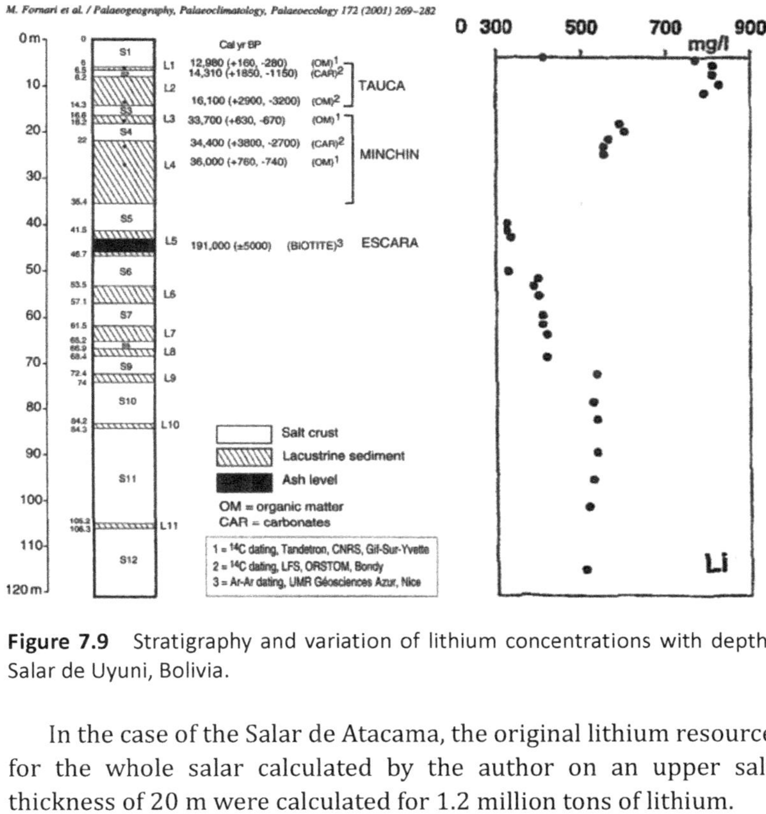

Figure 7.9 Stratigraphy and variation of lithium concentrations with depth, Salar de Uyuni, Bolivia.

In the case of the Salar de Atacama, the original lithium resource for the whole salar calculated by the author on an upper salt thickness of 20 m were calculated for 1.2 million tons of lithium.

Drilling by Minsal revealed that the salt porosity reached down to 40 meters and consequently updated the lithium and potassium resources of the Salar.

Drilling information shows that a major N–S fault transects the salar and the eastern side subsided resulting in salt accumulation reaching 1200 m while the salt thickness on the western side is 400 m thick. However, studies by Houston indicate that in mature salars, those composed mainly of salt (Atacama, Hombre Muerto, and Rincon), the porosity of the salt decreases because of compression, decreasing from about 12.5% near the surface to about 5% at 40 m and 0% at 100 m.

In lithium-containing brines, the sodium, potassium, magnesium, calcium, boron, and sulfate must be minimized to produce a lithium chloride solution suitable for lithium recovery as carbonate or chloride. Lithium carbonate is converted into many downstream chemicals and lithium chloride into lithium metal (although recently some companies are producing lithium metal from high-purity lithium hydroxide).

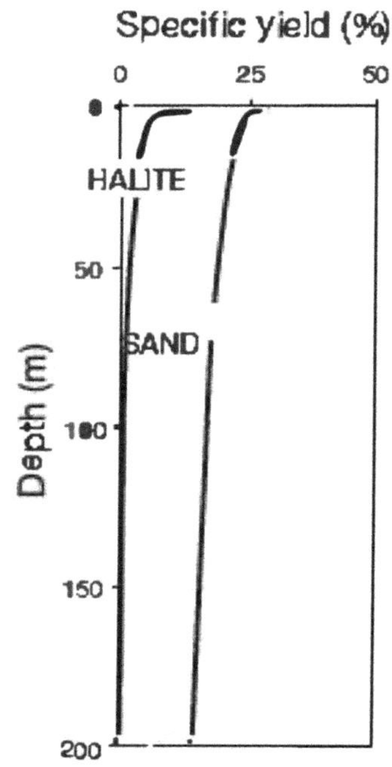

Figure 7.10 Decrease in salt porosity , Salar de Atacama, Chile (after Houston).

While brines are principally Na-SO_4-Cl systems, the geochemical environment has resulted in brines of materially different composition, both in the concentration of lithium and the presence of impurities. Thus, a single approach to lithium production is not applicable and purification of the brines is necessary, which will be discussed in Chapter 9.

Table 7.1 Compositional differences between various salars

Brine Deposit	Li	K	SO_4	Mg	Li	SO_4/K	SO_4/Mg	Mg/Li
Salar de Atacama	0.15	1.85	1.46	0.96	9.7	0.79	1.52	6.4
Hombre Muerto	0.07	0.6	0.9	0.21	12.9	0.67	4.29	3.0

Brine Deposit	Li	K	SO_4	Mg	Li	SO_4/K	SO_4/Mg	Mg/Li
Silver Peak Nevada	0.02	0.8	0.71	0.03	35.5	1.13	23.67	1.5
Rincon	0.033	0.62	1.1	0.28	33.3	0.56	3.93	8.5
Olaroz	0.066	0.55	1.7	0.19	25.8	0.32	8.95	2.9
Cauchari	0.051	0.42	1.57	0.15	30.8	0.27	10.83	2.8
Diablillos	0.05	0.21	0.75	0.55	15.0	0.28	1.36	11.0
Uyuni	0.045	1.2	1.4	1.07	31.1	0.86	1.31	23.8

To better characterize the different compositions of various lithium brine deposits, the Janecke diagram (Fig. 7.11) illustrates different compositions of the various brines, which ultimately control the chemistry and resultant salts during solar evaporation. The diagram summarizes the varied chemical compositions of different salars as a function of their magnesium, potassium, and sulfate concentrations. To be able to produce lithium chemicals, various brines will have to be modified so that none of the extraneous elements co-precipitate with lithium as complex potassium, magnesium, and sulfate salts, effectively removing it from the brine and rendering the process uneconomic.

Although the fundamental character of the salars is similar, a cursory examination of the three lithium-rich basins, which I will describe below, reveals great variability in size, surface character, stratigraphy, structure, and chemistry.

In an immature salar, there are several phases of high evaporation as well as salt deposition and dilution during pluvial episodes, causing dilution of the existing brines and deposition of sediments as evidenced in the stratigraphy of Clayton Valley, Salar de Cauchary, and the Salar de Uyuni.

Every deposit is unique because each different brine composition will result in a different process. Furthermore, conditions such as climate, reservoir hydrogeology, reserves, infrastructure, and freshwater availability will dictate whether a salar or Salt Lake is technically and economically exploitable. The final production process will depend on all these factors, which, as can be understood, is a complex task.

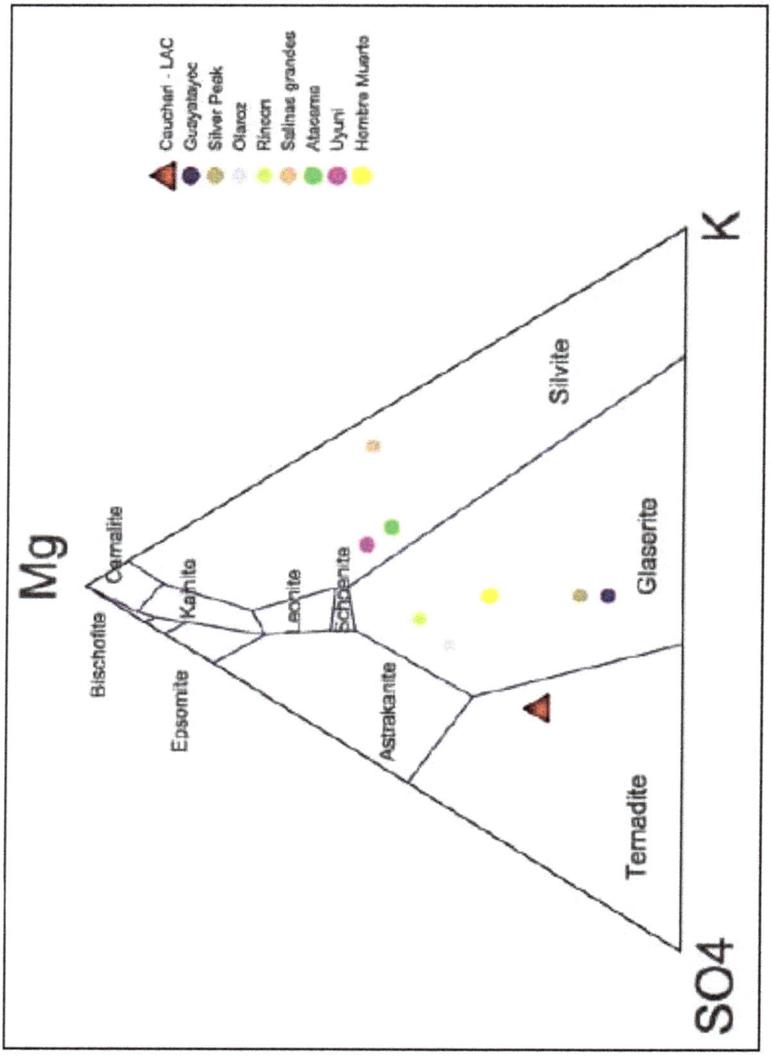

Figure 7.11 Janecke diagram illustrating compositions of various salars.

In general, the two elements of major economic interest in salt systems are lithium and potassium (bromine). To make a very rough estimation of their recoverability, the following parameters can be considered:

1. **Concentration:** The initial concentration of lithium and potassium in the brine is a fundamental requirement. This will define the concentration factor, the technical recovery approach, and the operational costs.
2. **K/Mg:** This ratio is very important for the recovery of potassium chloride. The higher the ratio, the more Sylvinite (KCl.NaCl) and less Carnallite ($KCl.MgCl_2.6H_20$) will crystallize in the solar ponds resulting in higher overall recovery.
3. **SO_4/Mg:** This ratio will indicate if the brine leads to a sulfate-potassium-magnesium (SulPoMag) salt (high SO_4/Mg), or a potassium sulfate salt produced from schoenite and/or kainite ($KCl.MgSO_4.2.75H_20$). However, if the potassium concentration is very low, it will generate a magnesium sulfate deposit.
4. **SO_4/K:** A high SO_4/K ratio will result in a potassium sulfate (SulPoMag) deposit produced from principally schoenite and/or kainite or potassium chloride deposit if the ratio is low.
5. **SO_4/Li:** A high SO_4/Li ratio will lead to a lithium sulfate deposit, or if the SO_4/Li ratio is low, a lithium chloride deposit.
6. **Mg/Li:** To produce a lithium chloride brine, this ratio is of importance, and it reflects how much magnesium must be removed to prevent unwanted lithium loss by complex salt precipitation.

Magnesium is one of the major factors impacting the economics of recovering lithium. In systems such as Silver Peak, Nevada, and the Salar de Cauchari, Argentina, the magnesium-to-lithium ratio is low and manageable and the removal of magnesium is economically feasible. In the salars of Bolivia and China, the magnesium is quite high and the magnesium-to-lithium ratios are as high as 26:1 and the removal of magnesium by simple lime addition is not economic. Therefore, magnesium rather than lithium is produced at the Great Salt Lake by the US Magnesium Corporation and potassium salts by GSL. For this reason, various attempts at producing lithium from these brines have been unsuccessful to date. In the case of the Salar de Atacama where the magnesium-to-lithium ratio is 6.5, Foote's analysis during the feasibility concluded that the addition of lime would be uneconomic. Exploration by the author located a high calcium brine source which solved the problem.

References

1. Lenntech.com, 2019, *Major Composition of Seawater.*
2. Israel Chemicals Company Profile December 2013, *BusinessWeek.*
3. Benson, L. V., Lund, S. P., Smoot, J. P., Rhode, D. E., and Spencer, R. J., 2011, The rise and fall of Lake Bonneville between 45 and 10.5 ka, Bonneville. *Quaternary International*, **235**, 57–69.
4. Behrens, P., 1967, Industrial processing of Great Salt Lake brines by Great Salt Lake Chemicals Corporation, in Gywnn, J. W. (ed.), *Great Salt Lake: A Scientific, Historical and Economic Review*, Utah Department of Natural Resources: Utah, p. 223.
5. Ogden Standard Examiner, 2/9/1969.
6. www.usmagnesium.com, "US Magnesium is nearing completion of the construction of a plant to produce lithium carbonate."
7. Ahfeld, F. E., 1972, *Geologia de Bolivia*, Los amigos del Libro: La Paz, Bolivia.
8. Ericksen, G. E., Vine, J. D., and Ballon, R. A., 1978, Composition and distribution of lithium-rich brines in Salar de Uyuni and nearby salars in southwestern Bolivia. *Energy*, **3**(3), 355–363.
9. Kunasz, I. A., 1975, Lithium raw materials, in *Industrial Minerals and Rocks* (4th Ed.), 791–803.
10. Garrett, D., 2004, *Handbook of Lithium and Natural Calcium Chloride: Their Deposits, Processing, Uses and Properties*, Elsevier Academic Press, https://doi.org/10.1016/B978-0-12-276152-2.X5035-X
11. Kunasz, I. A., 1970, *Geology and Geochemistry of the Lithium Deposit in Clayton Valley, Esmeralda County, Nevada*, Pennsylvania State University Thesis.
12. Ide, F. Y., and Kunasz, I. A., 1990, Origin of lithium in Salar de Atacama, Northern Chile, in Ericksen, G. E., Pinochet, M. T. C., and Reinemund, J. A. (eds.), *Geology of the Andes and its Relation to Hydrocarbon and Mineral Resources*, Vol. 11, Circum-Pacific Council for Energy and Mineral Resources: Houston, p. 165.

13. Munk, L. A., *et al.*, 2016, *Lithium Brines: A Global Perspective*, Geoscience World, https://doi.org/10.5382/Rev.18.14

14. Houston, J., Butcher, A., Ehren, P., Evans, K., and Godfrey, L., 2012, The Evaluation of brine prospects and the requirements for modifications to filing standards, *Economic Geology*, **106**, 1225–1139.

15. Risacher, F., and Fritz, B., 1991, Quaternary geochemical evolution of the salars of Uyuni and Coipasa, Central Altiplano, Bolivia, *Chemical Geology*, **90**(3–4), 211–231.

16. Kunasz, I. A., 1974, Lithium occurrence in the brines of Clayton Valley, Esmeralda County, Nevada. In Coogan, A. H. (ed.), *Fourth Symposium on Salt: Northern Ohio Geol. Soc.*, Vol. 1, pp. 57–66.

Chapter 8

Lithium Brine Exploration

Since lithium had become such a chemical star because of the promise of future electric cars and energy storage batteries, a plethora of newcomers have started exploring lithium. The Tesla lithium factory has had a big impact on the expansion of exploration in Nevada. At one time, there was even a lithium "highway" drawn from Reno to south-central Nevada, where the dry lakes of the Southwestern United States became a target, triggered by the occurrence of lithium-rich brines in the Clayton Valley in southern Nevada.

Most of the Nevada basins have formed because of the collision of the Pacific and the Continental plates. The resulting extension created structural depression which was subsequently filled with sediments and various potassium, lithium, magnesium, and other elements. By and large, the depressions are northeast-trending, are bounded by faults and are generally tilted to the east. This morphology gave rise to the term "basin and range" and as Maj. Clarence Dutton one of the early explorers colorfully noted, the map of Nevada looks like "*an army of caterpillars crawling northward out of Mexico*" [1].

A cursory view of the geological map of Nevada would lead one to suspect that many basins would be good candidates for the occurrence of a lithium deposit. However, that is not the case.

There are several conditions required for such a deposit:

- A source of lithium

The Lithium Legacy
Ihor Kunasz

ISBN 978-981-4968-31-7 (Hardcover), 978-1-003-37236-3 (eBook)
www.jennystanford.com

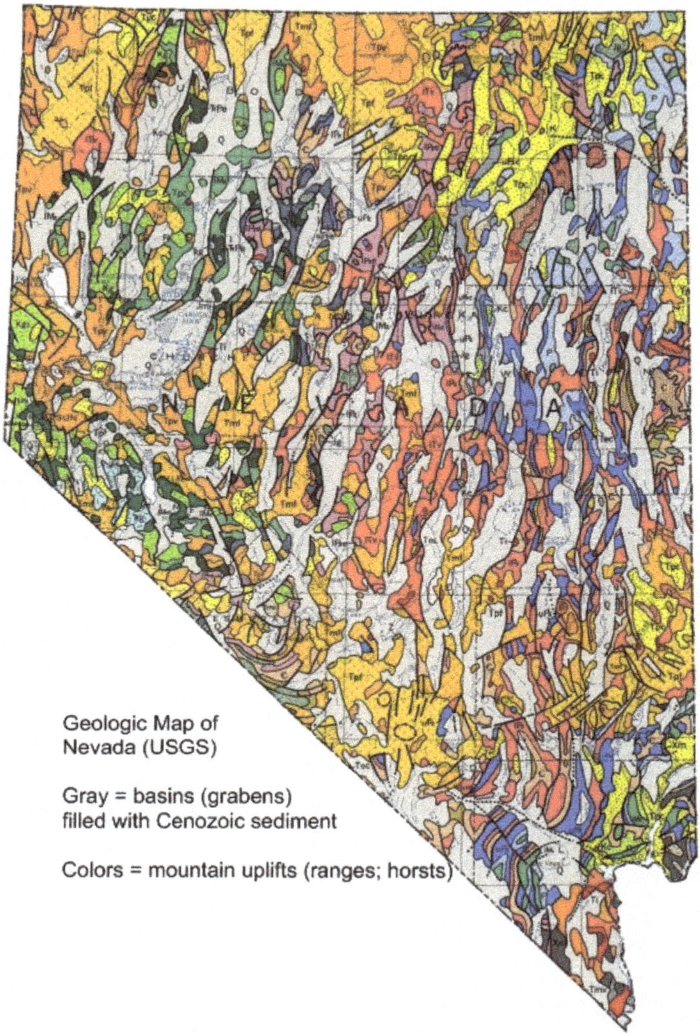

Figure 8.1 Geological map of Nevada exhibiting structural linearity.

- A closed structural depression (where lithium solutions are retained)
- An arid climate to concentrate the weak discharged lithium solutions (as witnessed by the presence of halite)
- An earlier history of lithium deposition

Brines are fluids contained in various sediments. They occur in environments that favor their accumulation in closed structural basins, into which lithium solutions (geothermal, leached) are concentrated by solar evaporation in the desert belts of the globe. When very rich solutions give rise to what Houston called mature salars (Salar de Atacama, Chile, Rincon, and Hombre Muerto Argentina), those that are primarily characterized by salt [2]. There are only a few of these.

Most other salars are characterized as immature because they consist of interbedded salt beds, deposited during dry periods and sediments were introduced during wetter pluvial periods (Clayton Valley, Nevada; Olaroz-Cauchari, Argentina; Uyuni, Bolivia).

One of the characteristics of the salars that is often overlooked is the recent alluvial fans. The white saline surface is the most recent evidence of the formation of the salar. As the basin subsides, alluvial fans originating during the pluvial periods encroach onto the salar and cover the peripheral portions of the basin. This is the case in the Clayton Valley basin. Originally, Foote Mineral Company claimed only the surface characterized by the white surface defined by salt minerals. Lithium Corporation promptly claimed the adjoining area covered by alluvial fans in the southeastern part of the northern basin of the basin. Later it abandoned those claims, and further exploration by Foote Mineral Company revealed that the salar extended under the alluvial fans, where, it turns out, the richest lithium brine had drained. A good exploration program must not focus exclusively on the salt flat but also on the peripheral geology to identify potential earlier lithium-sediments and laterally to take account of alluvium which may hide a portion of the salar.

In the case of the basin and range, only the Clayton Valley basin satisfy all four conditions. Many basins have been recently explored (Alkali Flat, Teels Marsh, Big Smoky Valley, Fish Lake Mud Lake, Railroad valley) and none have yet satisfied all these conditions.

The extensive exploration work of the United States Geological survey (USGS) has further documented the absence of any potential brine deposits. The suggestion that the lithium originated from halite dissolution leads to two conclusions: firstly, that the lack of halite in all basins indicates that the solutions entering practically all Nevada basins were so dilute that concentration never reached halite precipitation, (which, in the case of seawater, requires 90% of evaporation before halite precipitates) and is the driving mechanism for concentrating weak solutions. Secondly, the concept that lithium in Clayton Valley originates from the dissolution of halite is not very sustainable as lithium does not substitute any element in the sodium chloride structure, but would be present as enriched lithium brine in the porous halite [3].

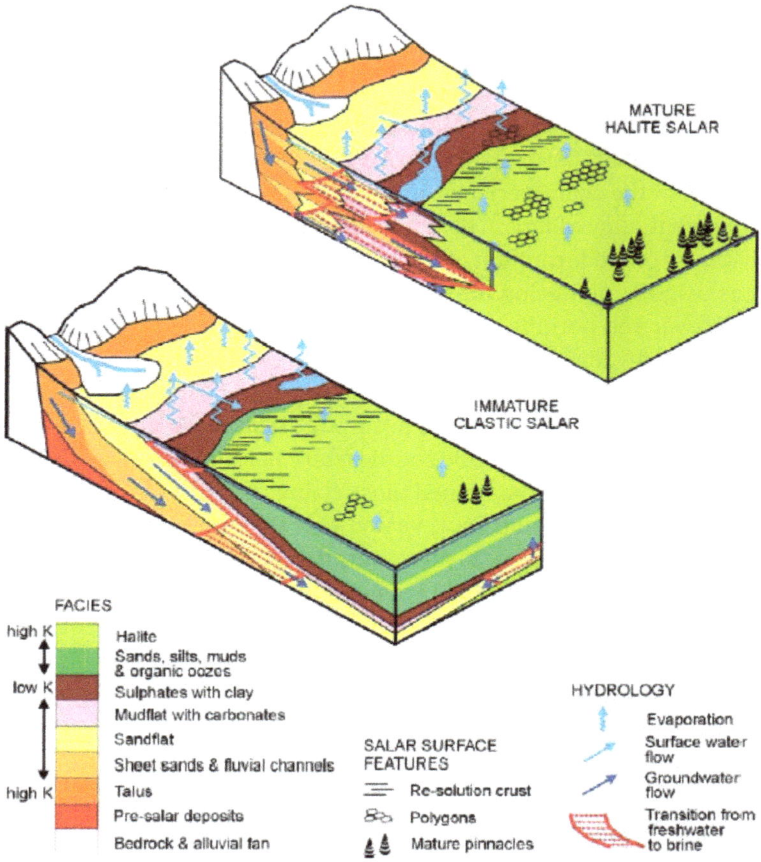

Figure 8.2 Mature and immature salar types.

While there are many dry lakes (or playas) in the Southwest United States, to date, none have met these conditions, and therefore, no economic lithium brine deposit has been identified although a great number of exploration programs have been undertaken.

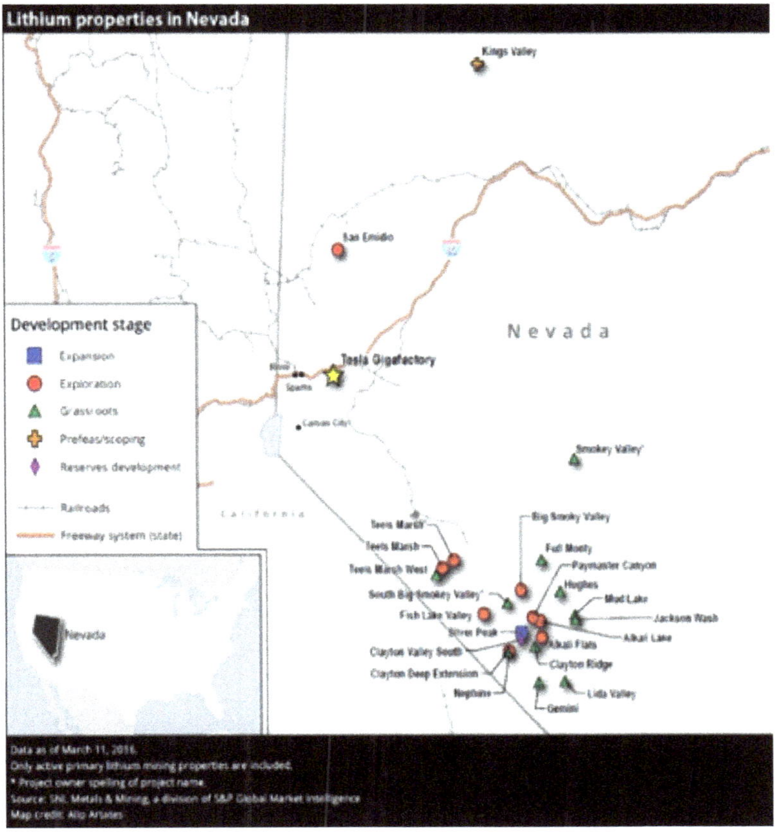

Figure 8.3 Location of exploration activities in Nevada.

This is supported by the extensive work conducted by the USGS [4]. At the Corning Symposium on lithium [5], members of the USGS, citing Hubbert's potential impact of the decline of petroleum reserves and the impetus to find alternate energy supplies, suggested that, according to the Argonne National Laboratory [6], there would be 18 million cars powered by lithium-sulfur batteries by the year 2000 and, therefore, the demand for lithium would grow exponentially. Forty years later, we know that this has not happened; global electric car production reached 2.1 million cars in 2019 [7]. The USGS accused lithium companies of dragging their feet in their lithium exploration efforts to locate new lithium deposits to satisfy the government automotive projections. One must remember that, after the U.S. government bought large quantities of lithium hydroxide to extract the lithium-6 isotope for the manufacture of hydrogen bombs, supply contracts ended after the atomic bomb was created. The U.S lithium industry was left with a large unused production capacity. In 1960 the sales volume of all lithium products by the industry was 25 million dollars. In 1970, the lithium industry was still recovering, and the lithium battery was in its infancy because there was no incentive by automotive companies to pursue this option. In fact, I clearly remember, at the symposium, the president of Foote Mineral Company, Evan Comer, stood, took out a purchase order book and invited the attendees to request all needed amounts of lithium, guaranteeing its prompt delivery to any customer. By 2017, lithium became a $2.8 billion industry and is predicted to grow to $5.9 billion [8].

In the 1970s, the government announced that a million electric cars would be operating in Washington, D.C., by 1980, a prediction that was never achieved. Based on this assumption, the USGS attempted to develop an exploration model designed to locate additional economic lithium brine deposits (which implied that Foote Mineral Company obviously missed!).

What the USGS did not know was that, after I was hired by Foote Mineral Company in 1970, my first assignment was to explore all the desert basins adjoining Clayton Valley and those of California. While my tools consisted of a six-foot gasoline-powered auger drill, the results of my exploration did not reveal any anomalous lithium values, which would have led Foote Mineral Company to acquire another target. Most interestingly, the low lithium concentration of 20 ppm Li in Death Valley, located directly south of Clayton Valley documented that it was isolated from Clayton Valley, otherwise, some anomalous lithium concentrations would have been recorded if the brine of Clayton Valley had leaked into Death Valley because of the great hydrostatic head differential between Clayton Valley (4200 ft asl) and Death Valley (below sea level). Since the Clayton Valley was an economic brine deposit, it made sense to try to identify the parameters that may help in locating additional lithium brines.

Figure 8.4 Author's early exploration activities, Death Valley, California.

The USGS started their exploration program in the periphery of Clayton Valley, an obvious target since the basin had been exploited for the past 40 years by Foote Mineral Company. Brine intercepted in wells drilled north and east of the brine deposit returned uneconomic values ranging from 55 ppm to 100 ppm Li in the northern sector and 110 to 190 ppm Li in the eastern sector of the basin. With such inconclusive values, the USGS requested permission to drill a well (CV-5) within the Foote Mineral Company well field brine basin proper, and not surprisingly yielded values of 280 ppm Li.

However, the USGS exploration program conducted in dry lakes adjoining Clayton Valley, resulted in disappointing values [9]. Fish lake to the west returned values of 7 ppm Li, Big Smoky to the north, 1.3 ppm Li (the USGS suggestion that the waters from Big Smoky leaked lithium into Clayton Valley was not supported by the data and, in fact, was challenged as early as 1906 by Dole) [10], Columbus Salt Marsh to the northwest returned 4 ppm Li, Alkali Flat to the east, 0.38–0.48 ppm Li. Other dry lakes in Nye County proved equally devoid of economic lithium economic values. Stonewall Flat, located further southeast of Clayton Valley registered only 0.16 ppm Li. Shallow auger holes drilled by Dajin Lithium Corp. at Teels Marsh returned 0.2 to 71 ppm Li. Fuse Cobalt acquired claims adjoining the Dajin project and their shallow sampling returned lithium values ranging from 8.9 to 104.5 ppm lithium (the latter collected downstream from thermal springs).

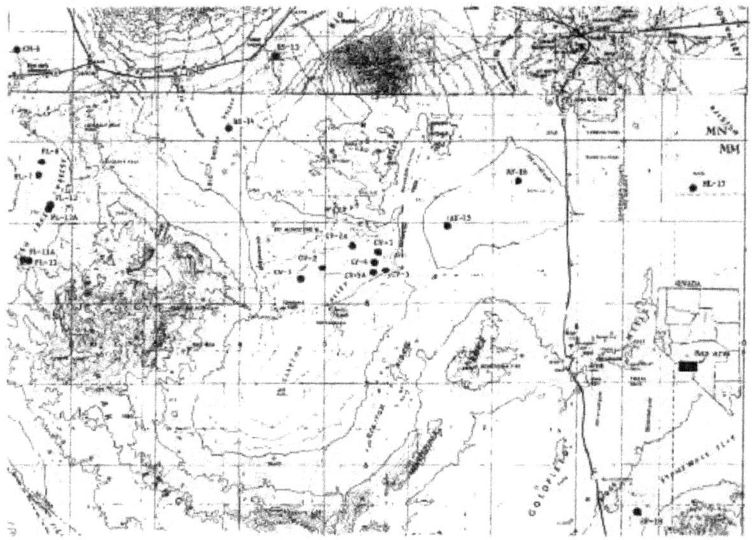

Figure 8.5 Location of the United States geological exploration activities in Nevada.

The USGS then conducted an intensive exploration of 23 dry lakes in California, very likely because of the presence of 50–70 ppm lithium in the brines of Searles Lake exploited for boron, sodium bicarbonate and sodium sulfate and had at one-time produced lithium carbonate. However, that campaign proved equally disappointing with most values in the low to mid-teens, with the best value from brines of Bristol Lake that reached 108 ppm Li, which has been recently reconfirmed by a 2020 study of Bristol, Cadiz, and Danby Lakes [11].

More than half of all brine samples drilled and sampled to depths of 504 feet by the USGS at 23 dry lakes returned values below 1 ppm Li (Alkali Flat, Broadwell, Coyote, Cuddeback, Dale, El Mirage, Emerson West, Emerson Central, Eureka, Ivanpah, Lucerne, Means, Melville, Soda). Others recorded higher values with depth especially at:

Ivanpah: 53–102

Lucerne:	35–102
Means:	29–102
Melville:	29–94
Soda:	22–126

Figure 8.6 Location of the United States geological survey exploration activities in California.

The low values are associated with higher conductivities, i.e., salinities), while the converse is recorded, indicating higher lithium values in intercepted fresh waters, a somewhat unexplained condition.

An exploration program in Panamint Valley by the Australian Battery Resource Company recorded mostly low lithium concentrations in shallow samples, with one near-surface sample analyzed at 120 ppm, probably the result of surface evaporative concentration. Panamint Valley is part of the system, which drained from Searles Lake and discharges into Death Valley. With low lithium values in Searles Lake (60 ppm) and 20 ppm in Death Valley, Panamint would not seem like a particularly attractive lithium exploration target.

8.1 Exploration Activities in Clayton Valley

Although the USGS did not come up with a promising exploration model, various junior exploration companies descended on Clayton Valley and recorded claims at the edges of the Foote Mineral Company (Albemarle) property and have conducted exploration with the hope of intersecting some economic lithium resources. Based on the USGS exploration data and the history of pumping, it is unlikely that economic lithium values would be intersected. Nevertheless, a plethora of "juniors" have swooped in and around Clayton Valley in the hopes to find a lithium "Eldorado."

- ACME Lithium Inc.: Fish Lake Valley (FLV) Claims, Dyer, Nevada
- Cruz Cobalt Corp.: Clayton Valley Lithium Prospect, Silver Peak, Nevada
- Cypress Development: Clayton Valley, Silver Peak, Nevada
  Dajin Lithium Corp.: Alkali Spring Valley, Alkali, Nevada
- Enertopia Corporation: Clayton Valley Lithium Project, Silver Peak, Nevada
- Ioneer: Rhyolite Ridge, Silver Peak, Nevada
- Jindalee Resources: Clayton North, Silver Peak, Nevada
- Marquee Resources: Clayton Valley Lithium, Silver Peak, Nevada
- Nevada Sunrise Gold Corp.: (Six lithium projects in Clayton Valley) Silver Peak, Nevada
- Nickel Rock Resources Inc.: Clayton Valley Lithium Project, Silver Peak, Nevada
- Noram Ventures: Clayton Valley Zeus Property, Silver Peak, Nevada

- Pure Energy Minerals: Clayton Valley Lithium Project, Silver Peak, Nevada
- Reedy Lagoon Corporation: Alkali Lake North Project, Alkali, Nevada
- Reedy Lagoon Corporation: Clayton Valley Project, Silver Peak, Nevada
- Scotch Creek Ventures: Highlands West Project, Silver Peak, Nevada
- Scotch Creek Ventures: Macallan East Project, Silver Peak, Nevada
- Sienna Resources: Clayton Valley Deep Basin Lithium Brine Project, Silver Peak, Nevada
- Spearmint Resources: Clayton Valley Lithium Prospects, Silver Peak, Nevada

While several companies have tried to identify economic brines, Cypress Development Corporation, on the other hand, has undertaken extensive testing of the Tertiary clays of the Esmeralda formation East of Angel Island, announcing planning to produce 27,400 tonnes of lithium carbonate equivalents (LCE) from clays averaging 1141 ppm Li based on a probable mineral reserve of 222,000,000 metric tons. Earlier attempts had been undertaken by Western Lithium, now Lithium Americas, in the McDermitt caldera, Northwestern Nevada. However, although the clays averaged 0.24% lithium, the economics did not support a lithium extraction facility, and the company ended up producing "hectatone" clay for the drilling industry [12]. The success of this product will be directly affected by the health of the drilling industry, which, under the new edict of President Biden to essentially shut-down oil and gas exploration, is questionable.

The optimism in the future of the electrification of cars and home storage units advanced by Tesla and other automotive companies has created a furor in lithium exploration in Nevada and South America. Junior Mining companies, mainly Canadian, acquire claims, conduct some preliminary exploration, and then sell them to other junior companies. So many exploration attempts have been initiated in the past few years that one would think that Nevada has become the lithium Klondike!

Nevada is unique as several basins have been created because of extension plate tectonics, creating several depressions, such as Clayton Valley, and all of which are believed to contain vast lithium resources. However, that is not the case because these dry lakes do not have all three required conditions (structural depression, arid climate (which leads to the concentrations of the lithium solutions as evidenced by halite deposits) and a geothermal lithium source. Unfortunately, these new explorers are mostly hard rock people who do not understand that these are fluid and not hard rock systems, where the elements (gold silver copper) are fixed. Soluble elements in brines are not fixed and change in composition as soon as they are pumped from the aquifer. They follow the exploration philosophy applicable to hard rock metal deposits, that is, saturate a dry lake with core drilling and analyze the lithium content of the pore liquids. Unfortunately, as illustrated by the excellent study of Houston et al. [2], while all sediments will be saturated with brine, not all sediments will yield the brine upon pumping, which is the only method to produce brines. While fine-grained sediments often comprise the bulk of sediments in a dry lake basin and can hold as much as 50% brine in their pores, their low permeability will never yield any brine. Exploration companies do not have a basic understanding of the genesis of lithium brine deposits nor of the hydrological conditions, which permit the economic extraction of lithium.

In many parts of Nevada, numerous dry lakes have been tested. Unfortunately, these start-up companies apply the 2012 Canadian Institute of Mining (CIM) NI 43-101 erroneous guidelines on brine resource reporting. Modelled on metals and industrial minerals deposits, these companies have saturated the dry lake basins with core holes. While the guidelines offer a rigorous reporting requirement for metals, brines are fluids and reserves, and resources cannot be modelled by the CIM guidelines. The various elements present in a brine deposit are not chemically bounded in a solid but occur in a dynamic fluid and will vary (decrease) with extraction by pumping. The following junior companies have staked placer claims in many dry lakes of Nevada:

- Altair International: Stonewall Lithium Project, Lida Junction, Nevada

- American Lithium: Tonopah Lithium Claims (TLC), Tonopah, Nevada
- American Pacific Borates Limited: Salt Wells Projects, Salt Wells, Nevada
- Belmont Resources: Kibby Basin Project, Tonopah, Nevada
- Caeneus Minerals: Columbus Marsh Project, Mina, Nevada
- Caeneus Minerals: New King Lithium Clay Project, Orovada, Nevada
- Caeneus Minerals: Rhodes Marsh Project, Mina, Nevada
- Dajin Lithium Corp.: Teels Marsh Lithium Project, Mina, Nevada
- Fuse Cobalt: Teels Marsh West Project, Mina, Nevada
- Great Basin Resources (Five lithium projects in Nevada)
- Iconic Minerals: Bonnie Claire Property, Beatty, Nevada
- Iconic Minerals: Smith Creek Valley Property, Austin, Nevada
- Jindalee Resources: McDermitt Lithium Project, McDermitt, Nevada
- Lithium Americas: Thacker Pass, Orovada, Nevada
- Lithium Corp.: Fish Lake Valley, Dyer, Nevada
- Lithium Corp.: San Emidio Valley, Empire, Nevada
- Macarthur Minerals: Reynolds Springs Lithium Project, Currant, Nevada
- Nickel Rock Resources Inc.: Black Rock Desert Lithium Project, Gerlach, Nevada
- Nickel Rock Resources Inc.: Dixie Valley Lithium Project, Dixie Valley, Nevada
- Scotch Creek Ventures: Cupz Project, Goldfield, Nevada

In California additional exploration is conducted by:

- American Pacific Borates Limited: Fort Cady Project, Newberry Springs, California
- Battery Mineral Resources: Panamint Valley, Trona, California
- Controlled Thermal Resources: Hell's Kitchen Lithium and Power Project, El Centro, California
- Pacific Imperial Mines: Eagle Mountain Lithium, Death Valley Junction, California

- Standard Lithium: Bristol Lake Mojave Project, Amboy, California

With the increased demand for lithium to supply the automotive and storage demands, it is expected exploration, whether successful or not, will increase. In 2020, Australian company Hawkstone Mining (now Arizona Lithium) announced a program of exploiting clays in northwest Arizona [13].

8.2 South American Brine Deposits

With the promise of an ever-expanding and successful lithium battery industry to satisfy the demand for car electrification, a plethora of "junior" exploration companies (Canadian and Australian) and investors (predominantly Chinese) have descended on the dry lakes of South America, principally Argentina, Bolivia, and Chile. Concurrently, many "specialists" keep on analyzing and rehashing *ad nauseum* the same information on the same salars.

In a sudden epiphany, the "Lithium Triangle" has been born. A figment of someone's imagination, the supposed triangle has nothing to do with the geological and geochemical reality of South American deposits. In fact, checking the "Google" Lithium Triangle, there are so many different triangles that they make one's head spin.

The areas where lithium occurs have been the products of the collision of the Nazca Pacific oceanic plate with the South American continental plate. Subduction caused the sediments to undergo melting, releasing metals as well as lithium, boron, and other valuable elements along vertical shears zones and discharging them onto endorheic basins, where under favorable climatic conditions of evaporation, economic brines in the Eastern Cordillera and the Puna. The tectonic maps do not support any "Lithium Triangle," but rather a rectangular zone parallels to the subduction zone and encompassing the Eastern Cordillera, where most of the lithium brine salars are located. Unfortunately, the Lithium Triangle has become a buzzword lacking any fundamental meaning. It should be more appropriately called the "Lithium Low-Value Stocks" Triangle since most of the companies are penny stocks.

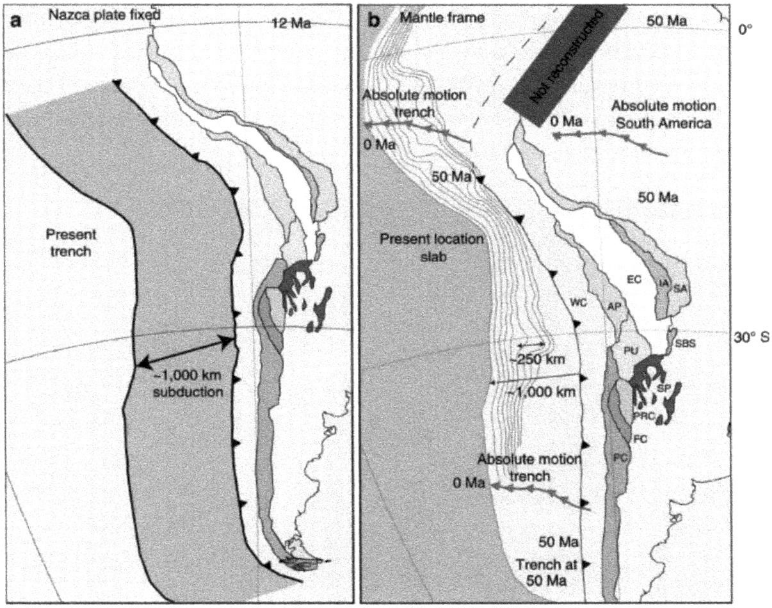

Figure 8.7 South American tectonic plates.

However, the oceanic plate did produce a belt of volcanic ignimbrite that extend on the western side of the Altiplano where many of the lithium-bearing salars occur.

There is no question that the presence of lithium in the South American salars of Argentina, Bolivia, and Chile is the result of subduction events that released hydrothermal solutions which ultimately discharged, as was the case of the Silver Peak Nevada system, into the evaporitic basin. It is interesting to note on the accompanying maps that the concentration of the lithium occurrences is located precisely in the Eastern Cordillera region and are absent to the north and the south, further supporting the sub-duction mechanism that introduced, lithium and boron into the Eastern Cordillera system.

While the major deposits, such as the Salar de Atacama, Chile, and Hombre Muerto, Argentina have been operating for decades, newcomers have not yet been able to develop economic deposits (except Salar de Olaroz and possibly Salar del Rincon in Argentina) but have "sniffed" (as many have done around the Albemarle operation in Clayton Valley or Livent at Hombre Muerto, Argentina) around existing operations, claiming peripheral areas hoping to identify economic lithium brines deposit. Reviewing historical data on salars, the exploration data by George Eriksen identified all the salars in Chile and Argentina.

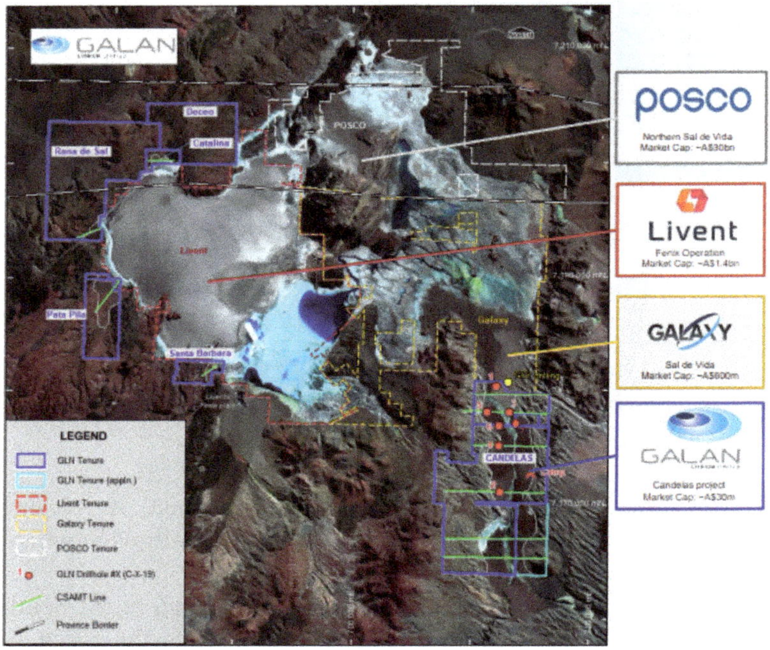

Figure 8.8 Salar del Hombre Muerto, Argentina showing the exploration activity surrounding the Livent existing lithium operation.

With so many salars identified in the Argentinian Altiplano, it is not a surprise that practically all the salars sampled and identified in the original report of Dr. Nicolli, have been partially explored. Whether they ever become economic is unclear at this time

Reuters reports that there are nearly 60 projects in various stages of development [14].

While extrapolation of 8760 km^2 has been quoted, it is not a clear conclusion that, while "the total area covered under tenements or mining leases or concessions will increase upon further research," it is not a guarantee that the existing nor the new concessions will hold economic lithium reserves, but certainly will provide much geological, geochemical, structural, and hydrological information on brine deposits.

Some of the projects are listed below [15]:

- Advantage Lithium: Cauchari
- Alpha Lithium: Salar Tolillar
- AIS Resources: Pocitos
- Alba Minerals: Pocitos
- Argentina Lithium and Energy Corp.: Arizaro
- Dajin LSC Lithium Resources Corp. Salinas Grandes: Pocitos
- Galaxy Resources Ltd.: Sal de Vida
- International Lithium Corp.: Mariana
- Lithium Energy Exploration Inc.: Laguna Caro, Antofalla North/South
- Lithium X Energy Corp.: Salar de Diablillos
- LSC Corp: Salar de Pozuelos
- Millenial Lithium Corp.: Pastos Grandes
- NeoLithium Corp.: 3Q: Catamarca
- NRG Metals Inc.: Hombre Muerto north
- Ultra Lithium Inc.: Salar Laguna Verde
- Wealth Minerals Ltd.: Laguna Verde

Furthermore, no data have been provided to suggest a volcanic/geothermal source that would have provided lithium solutions into a prior tertiary system (as was the case in Clayton Valley) to be recycled and reconcentrated into the present quaternary salars. Therefore, many salars have been identified but, as the bible says, few are chosen.

In conclusion, brine deposits are not gold, silver or copper deposits injected along weakness zones and exploration next to these deposits or along the geologic structures may lead to the discovery of new metals deposits. Brine deposits are liquids and, after years of pumping, the lithium concentrations at the periphery of the producing deposits decline and the chances of economic brines discovery are unrealistic.

In new regions, companies have essentially claimed any basin that shows some white crust and proceed to core drill in the hope of intersecting some rich lithium brine. Those efforts seem to be best illustrated by the Field Enterprises cartoon.

Figure 8.9 Cartoon illustrating the approach to lithium exploration by many hopeful junior exploration companies.

References

1. Dutton, C. E., 1880, *Report on the Geology of the High Plateaus of Utah*, U. S. Geographical and Geological Survey of the Rocky Mountain Region, Government Printing Office: Washington, DC.
2. Houston, J., Butcher, A. S., Ehren P., and Evans, K., 2011. The evaluation of brine prospects and the requirement for modifications to filing standards. *Economic Geology*, 106(7), 1225–1239.
3. Vine, J. D., 1980, Where on Earth is all the lithium? USGS Open file Report 80, 1234.
4. Davis, J. R., Friedman, I., and Gleason, J. D., 1986, Origin of the lithium-rich brine, Clayton Valley, Nevada, *Shorter Contributions to Isotope Research, USGS Bulletin* 1622, 131–138.
5. Penner, S. S. (ed.), 1977, *Lithium Needs and Resources*, Proceedings of a symposium held in Corning, New York, 12–14 October 1977, 1st Edition.
6. Chilenskas, A. A., Bernstein, G. J., and Ivins, R. O., 1976, Lithium requirements for high-energy lithium-aluminum-iron-sulfide batteries for load-leveling and electric vehicle applications, in *Lithium Resources and Requirements by the Year 2000*, Geological Survey Professional Paper 1005.
7. IEA, Global EV Outlook 2020. https://www.iea.org/search?q=Global%20EV%20Outlook%202022
8. Garside, M., 2020, Global lithium market value 2019–2027, www.statista.com/statistics/1088734/global-lithium-market-value
9. Pantea, M. P., and Asher-Bolinder, S., 1982, Lithologic log and lithium content of sediments drilled in Clayton Valley, Esmeralda County, Nevada. *USGS Open File Report* 82–415.

10. Dole, R. B., 1913, Exploration of salines in the Silver Peak Marsh, Nevada, *USGS Bulletin* 520-R, 1–18.
11. Rosen, M. R., Stillings, L. L., Kane, T., Campbell, K. M., Vitale, M., and Spanjers, R. (2020)., Li and Ca enrichment in the Bristol Dry Lake brine compared to brines from Cadiz and Danby Dry lakes, Barstow-Bristol Trough, California, USA, *Minerals,* **10**(3), 284.
12. Lithium Americas: Hectatone Subsidiary Enters Into Strategic Alliance with Tolsa, April 2016, www.lithiumamericas.com
13. Big Sandy Lithium Project, www. arizonalithium.com
14. www.reuters.com
15. Dickson, E., Lithium Triangle: South America's prospective. *Resource World*, www.resourceworld.com/lithium-triangle

Chapter 9

Techniques for Lithium Brine Resource Estimation

The high lithium demand forecasted for some high-tech applications and batteries for electric and hybrid-electric vehicles, in particular, has sparked increased exploration for lithium, particularly for lithium- and potassium-bearing brine deposits. Resource reporting for brines has generally followed the Canadian Institute of Mining (CIM) NI 43-101 guidelines formulated for hard rock deposits (metal and industrial minerals). While many requirements can be appropriately applied to brines, they still are fluids and require a specific analysis, leading to a distinct set of reporting requirements. Brines cannot be considered industrial minerals because the various elements occur as ions in a dynamic fluid, rather than being chemically bonded in a solid. Although chemicals are produced from brines, brines have no relation to industrial mineral deposits. This chapter proposes to analyze the recently released (November 1, 2012) document "CIM Best Practice Guidelines for Resource and Reserve Estimation for Lithium Brines," in light of the practical knowledge available from existing brine operations and recommends modifications to the guidelines, which if applied in their present format may result in overstating resources and reserves and potentially mislead investors [2].

The Lithium Legacy
Ihor Kunasz

ISBN 978-981-4968-31-7 (Hardcover), 978-1-003-37236-3 (eBook)
www.jennystanford.com

9.1 Introduction

The “best practice” document is a generally descriptive document, which does not fully address the methodology for brine resource or reserve definition. Lithium and other elements are contained in a fluid, and their identification should be the principal focus of a brine exploration program. The fundamental requirement in defining lithium brine resources and reserves in a salar is the identification and definition of those formations—the aquifers—that will yield brine in a usable quantity to a well, because brines can only be extracted by pumping wells. Fine-grained sediments will never, as required in the summary table (page 9), have “reasonable prospects of economic extraction,” and therefore need not be tested. However, the guidelines recommend the collection of solid samples and a high core recovery.

9.2 Analysis

While lithium-bearing fluids are present in the pores of various salar sediments, only those formations that will release brine when a pressure differential is applied (by a pumping well) should be investigated. In the CIM Guidelines, the term “aquifer” is used throughout the document, but no formal definition is provided. An aquifer is:

- A water (or brine)-bearing layer of rock, or of unconsolidated sediments, that will yield a usable quantity of water (or brine) to a well.

The document, instead, discusses “salar porosity and permeability and transmissivity on resource volume” (page 2),” and “salar brines are contained within a matrix” (page 4). This terminology is incorrectly applied to include all sediments because it can only be applied to those specific formations—the aquifers—that can yield brine upon pumping; they should not encompass the whole salar matrix, which includes fine-grained sediments that will not release brine upon pumping, and therefore, cannot be included in resource determination. The mechanism for “extracting” brine involves a well that is typically constructed using techniques similar to a water well. Pumping is the only method for brine recovery (sometimes trenching, in the upper portion of some mature salars). For this reason, an exploration program must first and foremost identify those formations that can produce brine and have reasonable prospects of economic extraction.

9.3 Resource versus Reserve

Once aquifers have been identified and their areal distribution established, a preliminary resource can be calculated using the following parameters:

(a) Aquifer areal extent
(b) Flow potential (to be later confirmed by pumping tests)
(c) Element concentration
(d) Hydraulic parameters

Refinement of resources (measured, indicated, etc.) should be based, as pointed out in the guidelines, on the density of the test drilling as well as the concentration grade cutoff and flow characteristics. Low flows and low concentrations should be categorized as indicated, while higher flow rates and high element concentration would be a measured resource. In no event should fine-grained sediments be included. Reserves should be defined by a tighter drill-hole pattern to better define the area, thickness, grain size (important in designing the gravel pack for production wells) and specific yield of the formation.

The continuity of the aquifer(s) is also important because it will predict long-term brine availability. As a production well impacts adjoining wells, the well-field design relies on the optimum interference between wells. If the interference pattern is areally extensive, it will confirm aquifer continuity and eliminate the need for more closely spaced exploration wells. Ultimately, this information will dictate the number of wells required to supply brine for a given chemical plant production capacity. In mature salars, numerical methods can be used to predict the drawdown of the brine under different pumping scenarios and to define the life of the reserves. The guidelines incorrectly recommend that "Resources may be defined as the product of three parameters:

1. Aquifer geometry
2. Specific yield
3. Element concentration."

Fine-grained sediments can be, indeed, characterized by area, specific yield, and element concentration. However, since those sediments will never yield brine, the calculation of a resource is not only irrelevant, but will generate an erroneously inflated resource number. In practice, the geologist logging the wells can readily identify permeable and impermeable sediments visually using the published information [3]. This is not explicit in the guidelines. The part of the porosity that may be drained under gravity is known as the specific yield or, sometimes, drainable porosity. To apply this parameter for resource calculation implies that:

(i) An aquifer(s) has been intercepted during exploration and
(ii) Brine will be available during pumping.

Figure 9.1 illustrates the relationship between various parameters [4]. Although clays and silts have a high porosity, they have a very low to negligible specific yield and high specific retention. Therefore, to recommend core drilling and sample collection of fine-grained sediments for porosity and permeability testing is unnecessary. Unfortunately, in some technical reports, resources have been calculated for all sediments in the salar. A review of geologic logs, however, indicates that clays and silty clays make up the bulk of basin sediments. The report includes extensive laboratory-tested specific yields, which were subsequently used to generate block models using kriging techniques. This is faulty since fine-grained sediments, while containing large amounts of brines in their pores, will never yield brine upon pumping. The reported resources are clearly overstated and invalid. Kriging techniques, if used, can only be applied to identified, known aquifers. The guidelines should recommend that detailed parameters be obtained only from identified aquifers, thus avoiding substantial costs in laboratory testing of fine-grained sediments and providing erroneously inflated lithium resources data.

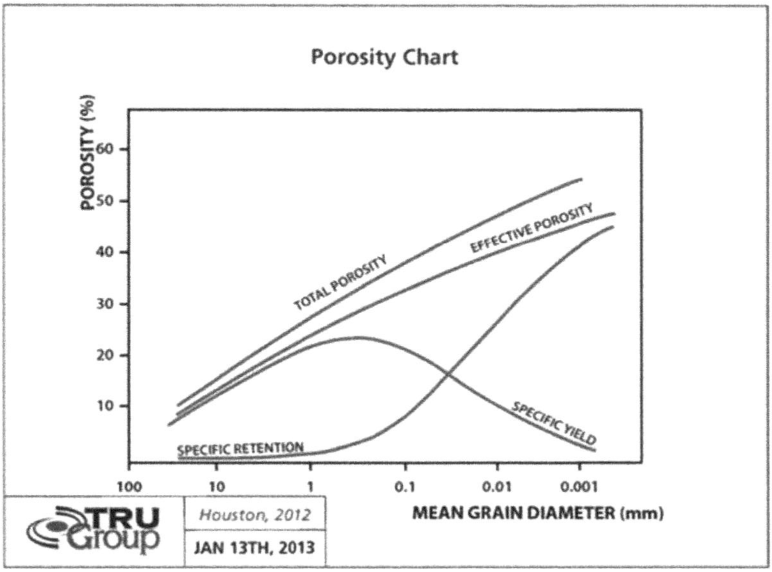

Figure 9.1 Porosity and (brine availability) as a function of sediment grain size.

9.4 Brine Exploration Methodology

The most efficient and proven exploration method for aquifer identification is the dual-wall reverse-circulation rotary (DWRC) drilling [6]. Designed by two Canadian drillers, the DWRC technology allows not only for the precise identification and recovery of the various sediments but also the collection of brine samples from producing zones by airlifting and identifying high- and low-flow intervals. Since the drill hole is cut with a close tolerance, excellent brine quality samples can be obtained from unconsolidated aquifers. At the Foote Mineral Company (Albemarle) Silver Peak Lithium Brine Operation in Nevada, hundreds of DWRC exploration holes and over 100 production wells have been drilled since the start of operations. The data provided not only accurate geological samples but also important hydrological information, which resulted in the determination of an empirical relation between the flows recorded during DWRC drilling and the expected production-well flows [1]. Additionally, practical experience indicates that sample contamination mentioned in the guidelines is not an issue.

It is remarkable that in several exploration programs, which used reverse-circulation drilling, no flow data have been reported. Therefore, in the first phase of exploration, the drilling procedures should focus on the identification of aquifers. While the DWRC drilling technology is well known for hard rock exploration, it appears that the committee members were not familiar with the importance of DWRC drilling as an exploration tool in salars. However, it should be incorporated in the CIM Best Practice Guidelines. The document states that qualified persons (QPs) should have "relevant experience in salar geology and brine processing." The requirement does not include experience in exploration of terriginous salars, in the identification and testing of brine aquifers, and in the design of production wells. Since there is only one such operation—Silver Peak, Nevada—it suggests that few have the overall expertise in recommending the appropriate exploration methodology, production of brine, and resource calculation from terriginous salars.

9.5 Recommendations

As stated in the introduction—the practices are likely to evolve and improve with time. There is practical experience in continental brines (Silver Peak, NV, with over 40 years of production records and Olaroz's detailed evaluation by Houston [4]) to recommend practical guidelines for the determination of brine resources. A brine deposit can only be qualified by the ability of aquifers to deliver brine to a pumping well, which satisfies the CIM requirement "to meet the definition of reasonable prospects of economic extraction."

Therefore, the first and foremost responsibility of the QP is to design a basic exploration program to identify the aquifers, which will eventually support production wells. The most practical and efficient exploration method is the DWRC drilling [5]. An exploration program should include the following steps:

- Secure proper permits and qualified contractor(s)
- Conduct shallow surface geochemistry to assess the distribution of lithium and other elements over the salar

- Analyze the geological and structural elements of the salar
- Conduct appropriate geophysical tests to define the geometry of the basin (basins are often asymmetric)
- Drill to identify aquifers by DWRC drilling (using the surface geochemical and geophysical data)
- Record the preliminary flow rates from the DWRC drilling
- Collect DWRC brine samples for chemical analysis
- Collect sediments for lithology definition and aquifer grain size analysis (important in designing production-well gravel packs)
- Define the areal distribution of the aquifers

9.6 Conclusion

The guidelines have attempted to modify the requirements for metals and industrial minerals, yet have retained much of their sampling methodology (core drilling, etc.). However, since the purpose of the document is to focus on lithium brines, there is a great need to improve the guidelines to truly reflect the most important requirement: the availability of sufficient brine and the identification and characterization of permeable formations, which have "reasonable prospects of economic extraction." This chapter will provide appropriate and practical guidelines for brine exploration to prospective exploration personnel and the analysis will offer positive recommendations for improving the existing best practice guidelines.

Acknowledgment

The author is grateful for the expertise and suggestions of E. Joel Christophersen, B.S. Geology, MBA, operations manager (former), Silver Peak, Nevada Brine Field.

References

1. Bell, R., 1984, *Advantages and Limitations of Dual-Wall Reverse Circulation Rotary Drilling for Ground Water in Unconsolidated*

Sediments, Professional Engineering degree, University of Wisconsin-Madison (unpublished).

2. CIM Estimation Best Practice Committee, 2012, *Best Practice Guidelines for Resource and Reserve Estimation for Lithium Brines.*
3. Heath, R. G., 1984, *Basic Ground-Water Hydrology*, U.S. Geological Survey Paper 2220, p. 13.
4. Houston, J., Butcher, A., Ehren, P., Evans, K., and Godfrey, L., 2012, The evaluation of brine prospects and the requirement for modifications to filing standards, *Economic Geology*, **106**, 1225–1239
5. Strauss, F., Story, S. L., and Melhorn, N., 1989, Applications of dual-wall reverse circulation drilling in ground water exploration and monitoring, *Groundwater Monitoring & Remediation*, **9**, 63–71.

Chapter 10

Lithium Production History

While many organizations and institutions have rehashed recent lithium production and demand statistics, the evolution and the history of the lithium industry have not been undertaken. The early pioneering names that propelled the lithium industry from its humble beginning to the present have been forgotten: Allied Chemicals, Maywood Chemicals, Foote Mineral Company, and Lithium Corporation of America. Today, some of these companies have been absorbed: Foote Mineral Company by Albemarle, Lithium Corporation of America by Livent.

Today, while there are several smaller producing countries, the lithium production industry is controlled by four important players: Soqumich (Chile), Albemarle (Chile, USA, Australia), and Livent (Argentina) from brines, and Talison (Australia) from lithium pegmatites. Lithium Minerals are also produced in Zimbabwe. Orocobre has started a substantial lithium brine operation in Argentina.

Production statistics are available in reports by the United States Bureau of Mines for the United States and the British Geological Survey for the Commonwealth countries. Production from the USSR is not published, although recent reports have identified several districts, including grades and resources. Recent publications on lithium in China have been published.

The Lithium Legacy
Ihor Kunasz

ISBN 978-981-4968-31-7 (Hardcover), 978-1-003-37236-3 (eBook)
www.jennystanford.com

It is interesting to note the evolution in the reported lithium units for lithium production. In the 1880s, newly discovered minerals were characterized by their oxide components. For example, the lithium mineral petalite, $LiAlSi_4O_{10}$ (as well as other minerals), was described as a combination of oxides, i.e., $Li_2O.Al_2O_3.Si_4O_6$ and lithium production was reported as units of Li_2O [1].

With the advent of the Atomic Energy Commission's production of the hydrogen bomb, which used the lithium 6 isotope extracted from lepidolite, production statistics were reported as lithium hydroxide ($LiOH.H_2O$) [2].

Following the termination of the Atomic Energy Commission (AEC) program and the increasing application of lithium carbonate in ceramics, production was reported as lithium carbonate (Li_2CO_3).

Today because of the increased development of lithium batteries using lithium metal, production is reported as units of elemental Li (or sometimes still as lithium carbonate equivalents (LCEs), a term which incorporates all lithium chemicals and minerals).

10.1 Early Years

The first extensive report on early lithium production, pricing, and technology in the United States was published by Arundale [2]. The lithium chapter in the *USBM Minerals Yearbook* includes foreign lithium activities. The most comprehensive report on United States' lithium production was published by Norton and Schlegel [3], which P. E. Landolt summarized [4].

The first recorded production in the United States was an estimated 30 tons of lithium minerals from an unknown source. Lithium production figures for the years prior to 1917 are poor; 1901–1903 was the period where more than 1000 tons of lithium minerals with a gross value exceeding $20,000 were produced [3]. At the turn of the 20th century, the best known and most productive lithium mineral producing centers were the zoned pegmatites operations at two major locations in the United States: California (Stewart mine) and South Dakota (Black Hills). In the 1950s, North Carolina (tin–spodumene belt) became the single supplier of lithium minerals (spodumene) and downstream chemicals.

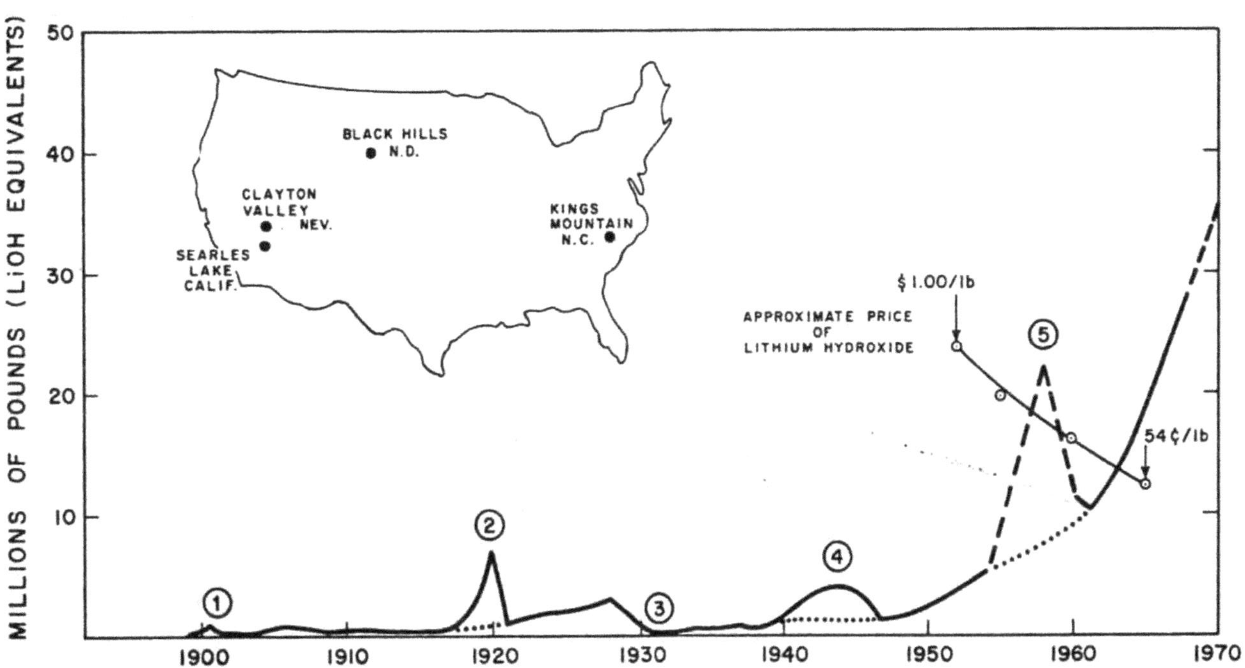

Figure 10.1 Curves showing lithium production in United States, as pounds of LiOH equivalents, and the price of lithium hydroxide in recent years. Index map shows locations of principal sources: each indicates point on curve that represent time of initial production from that source. Numbers refer to irregularities on production curve caused, in order, by (1) opening of lepidolite bodies at the Stewart mine, San Diego County, California. (Fig. 10.2); (2) renewed activity at the Stewart mine and the first use (during World War I) of lithium in alkaline storage batteries; (3) the economic depression; (4) use (during World War II) of large quantities of lithium hydroxide to inflate rescue apparatus, and in the development of multipurpose lithium greases; (5) lithium purchase program of the U. S. Atomic Energy commission.

During World War I (WWI), there was renewed activity in mining the lepidolite as lithium was used in alkaline storage batteries. In a patent application filed on May 10, 1907, Thomas Edison explained that adding two grams of lithium hydroxide to every 100 cc of electrolyte solution caused his battery's capacity to spike by 10% and extended the amount of time the battery could hold a charge by a "remarkable" amount. Today we know that the lithium hydroxide most likely helped avert some detrimental, unintended, chemical reactions that had been sapping away the battery's strength. Edison, however, had no clue because it worked, and he probably didn't care. Edison didn't build anything resembling a true lithium battery. Lithium was the salt in his stew. Today, we know that lithium is the best possible foundation for electrochemical energy storage. The universe hasn't given us anything better. Lithium, which is now used for purposes as diverse as treating bipolar disorder and strengthening aircraft frames, is one of the three primordial elements, created during the first minutes after the big bang. The lithium atoms in our laptops and cell phones are among the oldest pieces of matter in the universe. Composed of three neutrons, three protons, and three electrons, lithium is the third element on the periodic table, preceded only by hydrogen and helium. A metal, it is half the density of water and, in its elemental form, too volatile to exist in nature. Pure lithium is silvery-white and soft, like cold Camembert cheese, and must be stored in oil to prevent it from reacting with air or water. During WWII, lithium hydroxide was used to inflate rescue apparatus.

The industry was once dominated by two major U.S. producers, Foote Mineral Company and Lithium Corporation of America. Each produced spodumene and other mineral concentrates (feldspar and mica) and chemicals in North Carolina until the late 1990s. At the Salar de Atacama, the production of lithium carbonate as a by-product of potassium from the rich brines by the Sociedad Quimica y Minera resulted in a dramatic reduction in lithium carbonate prices and caused the shutdown of both U.S. pegmatite operations. Meanwhile, Australia, Canada, and Zimbabwe continued to supply lithium mineral concentrates for the ceramic and glass industry and other applications. Minor producers in Brazil, Portugal, Russia, and the People's Republic of China mine various lithium minerals. One new U.S. supplier TOXCO recycles lithium batteries.

When it was first recognized, lithium was an oddity that became an important commodity owing to its unusual properties. In 1854, in Germany, R. Bunsen and A. Matthiessen used an electrolytic process to prepare lithium as a free metal from molten lithium chloride. To become self-sufficient in lithium, Metallgesellschaft of Germany produced lithium carbonate, lithium chloride, and then lithium metal from zinnwaldite, a lithium-bearing mica mined in the Erzgebirge mountains of southeast Germany.

Not until WWII were the special properties of lithium compounds fully investigated and exploited. A compact, lightweight source of hydrogen was needed for use in emergency-signaling balloons. Lithium hydride was found to be ideal for this purpose. Lithium was also used in alkaline batteries in submarines. Later, greases containing lithium stearate were found to lubricate at both very high and very low temperatures. For the first time, the same grease could be used for multiple purposes over a wide range of operating conditions. With rocketry came the search for materials that could withstand the extreme temperatures of high-speed travel through the atmosphere. A ceramic composition containing lithium was developed that expanded very little and resisted cracking during rapid extreme temperature change. This lithium-containing material was the forerunner of modern glass-ceramic cookware that resists thermal cracking. Perhaps the most recognized application is Corning Ware, in which lithium imparts a negative coefficient of expansion when heated, which allows the ceramic to be used from refrigerator to oven without shattering. In 1953, the AEC required large amounts of lithium hydroxide from which the lithium-6 isotope was separated and reserved for use in producing the hydrogen bomb. For about 5 years, the government was the largest consumer of lithium. After the AEC contracts expired in 1960, the lithium industry, faced with vast overcapacity, sought desperately to develop some commercial markets. Though not an overnight success, it soon became a firmly established supplier to basic industries such as ceramics, lubrication, aluminum reduction, and pharmaceuticals. If certain technical issues are resolved, thermonuclear fusion, which requires lithium as the primary fuel, could solve much of the world's energy requirements. Today, even though lithium products are widely used in households, factories, and laboratories, lithium's presence often goes unrecognized. Lithium may be as close to the average person as a medicine chest, a television, a swimming pool, or a calculator. Lithium is found in minerals, clays, and brines in various parts of the world. High-grade lithium ores and brines are the present sources for all commercial lithium operations. Economical brine sources of lithium were rare until several salars in South America were discovered to contain significant deposits of lithium. Lithium was first produced from zinnwaldite in Germany. This was followed by the production of spodumene from the Black Hills of South Dakota, where log-sized spodumene crystals were mined. The WWII exploration for strategic elements (tin, tantalum, and others) resulted in the discovery of the pegmatite fields in North Carolina [5] where two major lithium mineral and chemical production centers developed. During the 1950s lepidolite from Southern Rhodesia (Zimbabwe) was imported for conversion to lithium hydroxide and separation of the lithium-6 isotope at a Texas plant for producing the hydrogen bomb. After the depletion of the lepidolite, a spodumene zone was outlined, resulting in the production of high-grade spodumene concentrates.

In the 1980s spodumene was identified during tantalum mining and exploration in Western Australia, resulting in the production of spodumene concentrates at Greenbushes.

10.1.1 Stewart Mine, Pala District, San Diego, California

The most important early lithium production came from the famous Stewart Mine. During the 1850s, ex-Gold Rush miners sought their fortunes in the waterways of the San Luis Rey River searching to find a new gold bonanza. Instead, as they panned and prospected up the river from the San Luis Rey mission church at Oceanside, they found a strange purple rock as well as Pink and Green Crystals in their gold pans. These curious 49's miners followed this water trail of bright color minerals to Pala and then up to their source on the Majestic Queen Mountain. They discovered a large pegmatite which, upon erosion, dumped gems into Salmon Creek. During the late 1800s, Chinese miners and laborers harvested vivid natural pink tourmaline crystals and sent them back to friends and family in China. Empress Dowager Cixi became interested in these gems and sent her officials to the mines to acquire more of the gems. These gems became popular among the rulers of the time, including the Romanov Czars and Queen Victoria. Even today in China, pink or red tourmaline gems are very popular and highly valued.

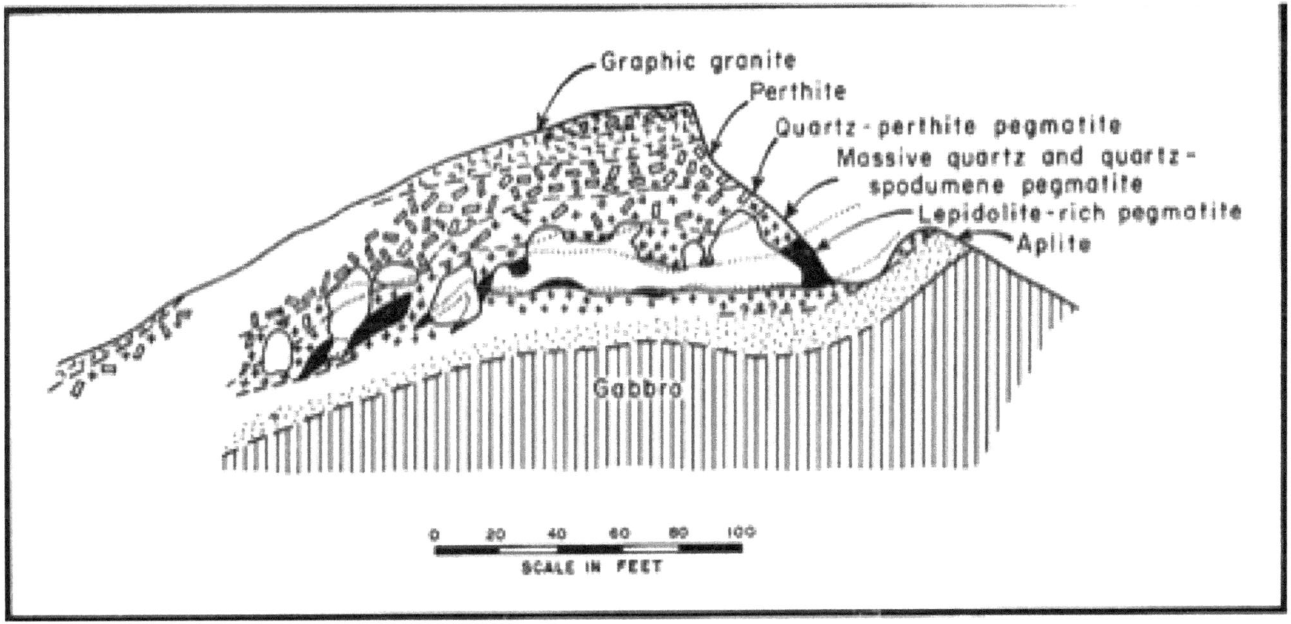

Figure 10.2 Generalized geologic cross section through zoned, lepidolite-bearing pegmatite dike at the Stewart Mine, San Diego County, California, modified from Jahns and Wright [27].

This pegmatite would later be called the "Stewart Lithia Mine," where lepidolite was mined as early as 1892, and soon became in 1902, the largest producer of lithium in the United States from its vast reserves of lepidolite ore, with some amblygonite. The greatest mining activity increased during WWI and reached a maximum of 11,696 tons in 1920 [3].

10.1.2 Black Hills, South Dakota

The most important pegmatite was the Etta Mine, Keystone, South Dakota. Spodumene was already produced in 1898 [6]. However, the 1920s were one of the most productive periods. Maywood Chemical mined spodumene at the Etta Mine where giant logs of spodumene were present. Lepidolite became important again in 1937 where it was mined from the Bob Ingersoll No. 1 pegmatite near Keystone. Lithium production reached a peak in WWII with a production of 84,800 units of lithia (1,696,000 pounds of Li_2O or 394 tons as Li).

10.1.3 Harding Mine, New Mexico

Some sporadic production of spodumene–lepidolite concentrate was produced in the 1920s from the Harding Mine, Taos County, New Mexico.

10.1.4 Wickenburg, Arizona

Mining of lithium minerals took place in Arizona at the Midnight Owl, also known as Lithia King near Wickenburg Arizona. The pegmatite belt extends from Lake Mead to Wickenburg for 250 miles. The southern part is the White Picacho District. Starting as a mica mill in 1947, the mine produced 62 tons of amblygonite—the lithium phosphate mineral—and 13 tons of beryl, with some columbite–tantalite concentrates. The mine procured minerals for the Maywood Chemical company and Foote Mineral Company [7].

10.2 The Atomic Energy Commission Years

Following WWII, the U.S. government began experimenting with hydrogen bombs. Large quantities of lepidolite were imported from Southern Rhodesia (Zimbabwe) from the Bikita pegmatites fields. Shipped to San Antonio Texas, the lepidolite was converted to lithium hydroxide monohydrate from which the lithium 6 isotope was separated. In 1952, the United States detonated the world's first thermonuclear weapon, the hydrogen bomb, on Eniwetok Atoll in the Pacific. The program was abandoned after the creation of the atomic fission bomb developed in 1945 at the Alamos National Laboratory, New Mexico. The thermonuclear capabilities of the hydrogen bomb are being evaluated as a potential source of energy if the initial reaction temperature can be contained.

In Germany, too, the possibilities of unexpected behavior were not lost on the scientists. When blockade imposed during WWI interrupted the supply of tin, Metallgesellschaft metallurgists alloyed lead and lithium. "Bahn metals" were used as a substitute for the bearing parts of railroad cars. Although the results seemed not to have been overwhelming, it did assume special importance as a pioneering project, and it likely stimulated more successful investigations into alloys of lithium with other metals (such as aluminum). After dealing in titanium (welding rods) and zirconium (radio and televisions), Foote Mineral Company became involved in lithium and became a leading producer of lithium compounds and lithium ores. A 1955 unpublished report "Lithium Ore Survey" summarizes in detail the knowledge Foote Mineral Company had at that time on all known lithium ore sources in the world.

Foote Mineral Company produced spodumene concentrate from its Kings Mountain mine and processed it into lithium compounds first at the Sunbright, VA, plant [8].

In 1953, Foote Mineral Company built a plant located at Sunbright, VA, because the chemical product was lithium hydroxide and the process to convert spodumene into lithium hydroxide monohydrate required the availability of limestone necessary to mix with the spodumene brought in the Kings Mountain mine. The spodumene–limestone mixture was in a rotary kiln at a temperature of 1030°F. After cooling, the clinker was leached with water to produce an impure lithium hydroxide solution, which was then fed to a vacuum crystallizer to produce a pure lithium hydroxide product [9]. Unfortunately, the residual sodium hydroxide (with some lithium) solution was discarded. The Sunbright area was underlain by limestone, with many sinkholes. These were used as a dumpsite for the waste. Unfortunately, connections between sinkholes, and the groundwater contaminated some of the private wells and the company agreed to connect the affected households to city water. After Foote Mineral Company bought the mill and mine from Solvay Process Co. in 1951, it moved all its lithium mining and processing operations to Kings Mountain, North Carolina, and began producing lithium carbonate. Spodumene has a theoretical lithium content of 8.03% Li_2O, but impurities such as iron and other elements reduce the lithium concentration in spodumene crystals. The pegmatite contains about 20% spodumene and has an average grade of 1.5% lithium oxide [10].

Mineral Content (% Wt)	
Spodumene	15-25
Potash feldspar	12-15
Soda feldspar	28-33
Quartz	25-35
Muscovite and other associated minerals	5-15

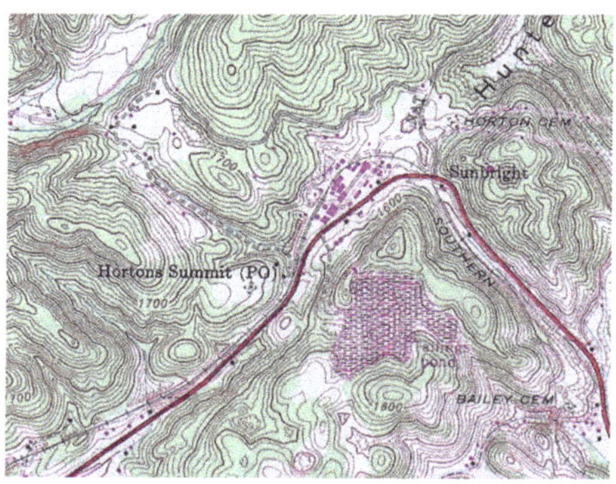

Figure 10.3 Sunbright, Virginia Foote Mineral Company lithium hydroxide plant.

The Foote Mineral Company spodumene pegmatite has been well-defined by drilling. Subsequent mining has confirmed the geometry of the pegmatite and a 30,000,000 tons resource of which 22,300,000 tons are in the reserved category. Although mining was discontinued in 1997, the reserves are sufficient to reopen the mining operation without major capital requirements, should an ever-expanding demand for lithium batteries for the automotive and energy storage industries materializes.

FOOTE MINERAL PEGMATITE CROSS-SECTION

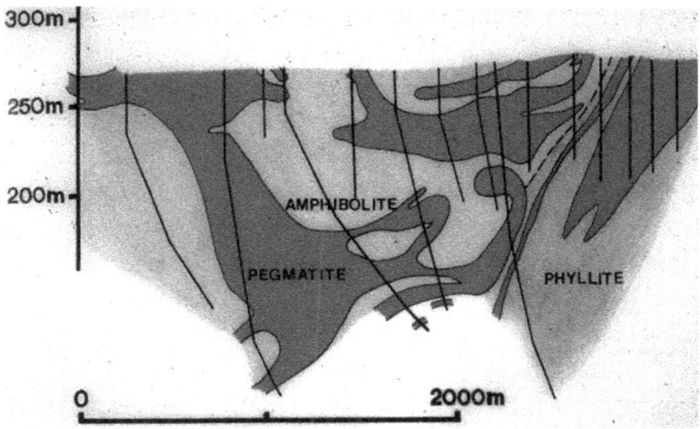

Figure 10.4 Generalized cross section—Foote Mineral Mine, Kings Mountain North Carolina.

Following the opening of the Foote Mineral Company Kings Mountain, North Carolina deposit, Lithium Corporation of America acquired the Hallman–Beam property north of Kings Mountain and outlined sufficient spodumene resources to start a parallel lithium operation. According to Evans, who conducted additional drilling to define the pegmatite ore bodies, the resources were estimated at 44,000,000 tons, of which 33,750,000 were categorized as reserves [11].

10.2.1 Lithium Corporation of America

From 1948 to 1950, the Lithium Corporation of America used a sink-float process to concentrate spodumene from the Edison mine near Keystone, South Dakota. In 1952, this company completed a spodumene flotation plant at Hill City, South Dakota, to process ore from the nearby Mateen pegmatite and Beecher No. 2 pegmatite, about 15 miles to the south. The Black Hills Tin Co. resumed mining at Tinton, but work came to a halt when the mill burned in November 1953.

By the 1950s, the lithium industry became totally concentrated in the tin–spodumene belt when Lithium Corporation of America joined in the mining of spodumene in acquiring the Hallman–Beam property some 10 miles north of Kings Mountain along the same tin–spodumene trend. Foote Mineral Company and Lithium Corporation of America fully controlled the lithium industry until the latter part of the 20th century when a substantial portion of the lithium chemicals production started at the Salar de Atacama, where two entities controlled the brine resources of the salar: Sociedad Chilena de Litio, Ltda. (SCL, a joint venture between Foote Mineral Company and the Chilean government), controlling the southern part of the salar and Sociedad Quimica y Minera (SQM) started operation the northern sector of the Salar de Atacama.

Today, a great quantity of lithium carbonate, which is the basis of various downstream lithium chemicals such as lithium hydroxide and lithium chloride, comes from the brines of the Salar de Atacama, Chile; Clayton Valley, Nevada; United States; and the Salar del Hombre Muerto, Argentina; and the recent Orocobre Operation at the Salar de Olaroz in Argentina. The Argentinean Salar del Rincon has been evaluated but as of now has not contributed substantial quantities of lithium carbonate. The Salar de Uyuni in Bolivia, while containing the largest lithium resource in the world, while unsuccessfully evaluating the production of lithium for the past 20 years, faces negative chemical (high magnesium) as well as political issues. Chinese researchers have been trying to resolve the unfavorable chemistry (mostly high magnesium) of their brine deposits in the Qinghai plateau where potassium salts are produced at this time. A recent announcement suggests that magnesium salts and possibly magnesium metals might be produced. Although intensive efforts have been made to produce their own lithium chemicals, China's lithium carbonate and metal production are sourced from Australian spodumene (Greenbushes).

10.2.2 Canada

In the 1950s, probably because of the AEC program, Canada tried to enter the lithium field by mining spodumene by the Quebec Lithium Corporation, which produced a peak of 42,000 MT of spodumene concentrates in 1957. Production declined to about 10,000 tons of spodumene concentrate operation ceased in 1968. While the termination is attributed partly to strikes, it is more likely that the start of the lithium brine operation by Foote Mineral Company at Silver Peak Nevada was, in addition to the long-term spodumene concentrate supply to Lithium Corporation, the real reason for ceasing the operation. I remember a comment by Evan P. Comer, vice-president of Foote Mineral Company that, after the president of Quebec Lithium Corporation visit the brine operation, it became clear that with the capacity of the existing lithium producers and the cheaper source from brines, that Quebec Lithium Corporation could not survive.

Low-iron spodumene and lepidolite for the glass and ceramics industries were produced during 1971–1972 by the Tantalum Mining Corporation at Bernic Lake. Besides spodumene and lepidolite, the pegmatite ore body also contains the cesium mineral, pollucite, and beryllium as well as tantalum. Tantalum Mining Corp. started production of tantalite in 1969 and announced plans to install a mill of capacity 350 tons/day to produce low-iron spodumene concentrates. The company is owned by the Goldfield Corporation (60%) and Chemalloy Materials Ltd. (40%). The Bernic Lake deposits are said to amount to about 7 MT of lithium minerals, assaying about 2.76% Li_2O. The spodumene is apparently an exsolution of petalite, having a very low iron content of 0.08% Fe_2O_3, and can be used directly in ceramic applications.

10.3 Lithium Production Today

For many decades, pegmatites were the sole source of lithium. With the discovery of lithium in Clayton Valley, Nevada, brines, especially in Chile (Albemarle and Soquimich) and Argentina (Livent and Orocobre) have replaced the pegmatites as their production costs were substantially lower. Today, the Greenbushes, Australia pegmatite body has taken the leadership in lithium production from spodumene.

While today's brine production has diminished after 50 years of operation, the Clayton Valley deposit merits a discussion as the geology and geochemistry have served as a model for many similar immature salar type operations.

10.3.1 Albemarle Silver Peak Nevada Process, Clayton Valley, Nevada

The project includes 4000 acres of solar evaporation ponds built on natural clay. At full production, some 40 wells pumped 11 million gallons of brine. During evaporation of any brine, all elements concentrate. According to the phase diagram, sodium chloride precipitates first, followed by sylvinite as a mixture of sodium and potassium chlorides. A sodium chloride base is deposited to a thickness of 12 or so inches during the first year of evaporation to provide a solid base for heavy equipment to operate to remove the sodium chloride layer, which accumulates annually to make room for the next season's accumulation. Otherwise, as salts precipitate, the dikes that enclose the ponds would have to be built up. Which option is chosen depends on the costs associated with slat removal versus dike construction. The sylvinite salts are harvested and deposited in a separate stockpile for further potential potassium chloride recovery. However, an early economic evaluation indicated that the potassium content of the sylvinite stockpile was too low to justify the costs of a separation plant.

Figure 10.5 Foote Mineral Company (today Albemarle) Pond Operations, Silver Peak, Nevada.

Continued evaporation of the brine increases the concentration of not only lithium but also magnesium, which results in the precipitation of complex potassium-magnesium-lithium salts and the loss of lithium. Consequently, the magnesium (and sulfate) must be removed for the lithium to be further concentrated as lithium chloride. In the case of Silver Peak, the ratio of magnesium to lithium is relatively low at 1.5, and it is economical to mix the pond brine with hydrated lime—$Ca(OH)_2$. The calcium reacts with the sulfate and the magnesium with the hydroxyl ion, resulting in a slimy mixture of calcium sulfate and magnesium hydroxide, which is pumped into a separate pond, where the mixture settles by natural decantation. The clear solution is moved to the next evaporation ponds and the settled mixture pumped out and disposed off in the salar.

From a starting concentration of 250 ppm Li, the brine reaches a concentration of 4000 ppm Li after a year and a half of evaporation. It is then pumped to the chemical plant where the remaining impurities are removed and purified lithium chloride solution is reacted with sodium carbonate, precipitating lithium carbonate, which is dried after filtration and stockpiled for shipping.

10.3.2 Albemarle Salar de Atacama Process

At the Salar de Atacama, Foote Mineral company has to tackle the high magnesium-to-lithium ratio. Following the signing of the feasibility study, exploration defined the bulk of the reserves over the whole salar. Because of the extreme ruggedness of the surface, the program had to be conducted by helicopter resources. The highly porous and permeable upper 20 meters of the crust indicated that high flow brine wells could be expected to support the pond operations. Based on the reserve evaluation, the Foote Mineral Company was granted enough area to guarantee a cumulative production of 200,000 MT Li. At the early stages of the feasibility, Foote Mineral Company had only the Silver Peak process in treating the magnesium and it was evaluated for application to the salar brines as well. Because of the high magnesium, the Silver Peak process proved to be uneconomic. Some other sources of calcium had to be identified. An evaluation of the geology of the southern part of the salar, which separates the salar into two subsectors, indicated that the Cordon Chilquinchoro is composed, in part, of gypsum and lacustrine carbonate rocks, which were made up of calcium and magnesium carbonates. Assuming the carbonate rocks had eroded and weathered and created the alluvial fans bordering the Cordon, the author recommended an exploration program to test this concept. The drilling program was successful as not only high-calcium but high-lithium brines as well were intercepted. This calcium brine soled the pond phase chemistry allowing the brine to be concentrated as a chloride and not lost as mixed precipitate. The availability of the calcium brine resulted in the construction of ponds sufficiently large to produce 12 million pounds of lithium carbonate as well as 40,000 tons of potassium chloride recovered from the sylvinite pond. The final brine, concentrated to 6000 ppm Li, is trucked to the La Negra production plant, located east of the city of Antofagasta to lithium carbonate.

Figure 10.6 1988 and 2022 Albemarle pond systems, illustrating the substantial growth in lithium brine production.

Albemarle acquired the interest of Rockwood Holdings in 2015, including the SCL salar as well as the Silver Peak lithium operations. In 2017, Albemarle inaugurated a second La Negra Plant dedicated to the production of battery-grade lithium carbonate with a capacity of 44,000 tons/year. In 2018, Albemarle brought a lithium chloride plant on stream.

With an ever-increasing demand for lithium chemicals and lithium metal for batteries, production increased from the initial 12 million pounds in 1988 to more than 11 times to 60,000 tons in 2020. In the spring of 2018, Albemarle received permission to increase their lithium mining there from 60,000 to 80,000 tons to up to 140,000 tons annually [12]. The recent map supplied by Albemarle illustrates not only the enormous increase in pond area but also both the number of high-calcium and high-sulfate wells (green dots). Quite a change from the original two high-calcium and one high-sulfate wells!

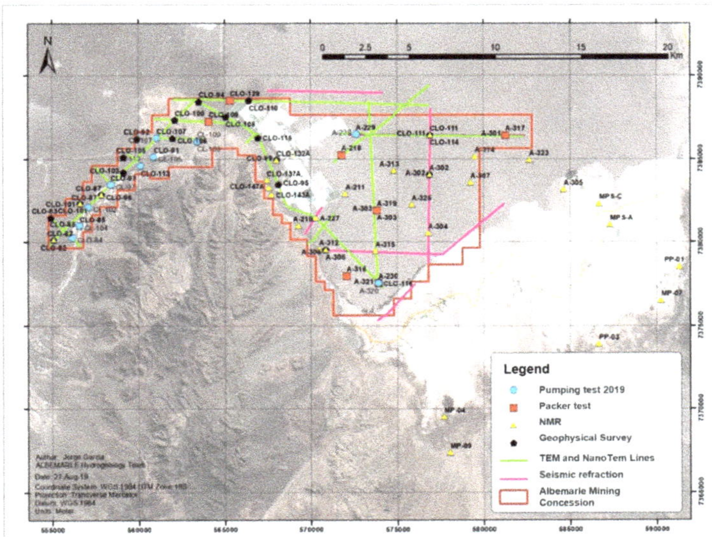

Figure 10.7 Location of the Albemarle well productions and pond system.

10.3.3 Salar de Atacama, Soquimich (SQM)

After Foote Mineral signed the contract for the feasibility study, the remainder stayed undeveloped and several companies, including Great Salt Lake Minerals, Lithium Corporation of America, MINSAL attempted to gain control.

In 1986, MINSAL was created between CORFO (25%), Amax (63.75%), and Molymet (11.25%). Under the guidance of Keith Evans, MINSAL undertook an extensive campaign of drilling 12 rotary holes, 34 core holes, and pumping tests at four locations to define the reserves of lithium and potassium more accurately. Results of the exploration indicated potassium (K) reserves of 24 million MT and lithium (Li) reserves of 1.67 million MT.

In 1989, SQM acquired the interest of Amax and Molymet and controlled the bulk of the 27,484 mineral tenements in the northern part of the salar. SQM was formed in 1968 through a joint venture between Compania Salitrera Anglo Lautaro SA ("Anglo Lautaro") and CORFO. In 1971, Anglo Lautaro sold all its shares to CORFO, and the company was wholly owned by the Chilean Government. In 1983, CORFO began a process of privatization by selling shares to the public and subsequently listing such shares on the Santiago Stock Exchange. By 1988, all shares were publicly owned. In 1995, SQM acquires total control of the northern claims by buying out the interests of CORFO, gaining control of 16,482 hectares and the rights to produce 180,100 MT of lithium until 2030.

Why was SQM interested in the brines of the Salar de Atacama? The nitrate operation in Coya Sur and Maria Elena (opened in 1926 after Guggenheim acquired the rights to the nitrate fields) required potassium chloride to produce the potassium nitrate fertilizers.

Since Chile did not possess a source of potassium chloride necessary to produce potassium nitrate, SQM had to import potash from Potash Corporation of Canada—an expensive cost component. SQM needed an alternate cheaper source. In the early years of production, SCL was producing about 30,000 tons/yr potassium chloride as a by-product from its lithium brine operation and was selling it to SQM. However, those quantities were not sufficient to satisfy the volume of potassium nitrate production.

As SQM did not have the advantage of high-calcium brines that Albemarle controlled in the southern part of their claims to control the brine chemistry, the company was able to pump two types of brines into separate solar pond systems: addition SQM operates a dedicated lithium pond system. In addition, SQM operates three production systems:

- potassium chloride ponds [high potassium (2.8%) and a low sulfate (0.35%)]
- potassium sulfate ponds [high sulfate (2%) and low potassium (1.7%)]
- potassium chloride—lithium ponds.

SQM started pumping rich potassium and lithium brines north of the SCL claims into a series of solar evaporation ponds, to produce 500,000 MT of potassium chloride, a portion of which is shipped to the nitrate operations, replacing the imported from Potassium Corporation of Saskatchewan, and exporting the excess.

Additionally, a zone of high sulfate in the center of the salar provided the opportunity to produce potassium sulfate, a fertilizer preferred over potassium chloride by some plants (tobacco, potatoes).

In 1997, SQM decided to produce lithium carbonate. After the precipitation of potassium chloride in the first set of ponds, the residual brine is enriched by a factor of 3. Since it could be considered a "free" brine, the costs of production being assigned to potassium chloride, SQM was able to offer lithium carbonate at half the commercial price, and in effect, resulted in the closing of the two North Carolina lithium carbonate produced from their spodumene properties.

The pond systems precipitate halite (NaCl—sodium chloride), sylvinite (NaCl–KCl,) and carnallite ($KMgCl_3 \cdot 6H_2O$—a hydrated mix of potassium and magnesium chlorides). These salts are further treated to produce potassium chloride, potassium sulfate, and the residual brine, lithium carbonate.

Figure 10.8 Soquimich early pond system.

Potassium chloride is used in the production of potassium nitrate. Production of SQM's own supplies of potassium chloride results in substantial raw material cost savings in the production of nitrate fertilizers at the Maria Elena, Nueva Victoria, and Pampa Blanca plants.

To produce potassium chloride, brines from the southern part of SQM Salar de Atacama claims, where the highest potassium (and low sulfate) concentrations exist, are pumped to the southernmost solar evaporation ponds system. After evaporation, the sodium and potassium salts are harvested and sent for recovery of potassium chloride by grinding the mixed halite–sylvite salt, flotation, and filtering process. Carnallite salts are harvested and sent for treatment at one of the cold leach plants where magnesium is leached, and additional potassium chloride is produced. SQM reports that the current production capacity more than up to 2.6 million MT/yr of potassium salts.

Figure 10.9 Areal view of the Soquimich pond system.

Following sequential evaporation, part of lithium chloride brine is transported to a plant in the Salar de Carmen, located east of Antofagasta, where it is processed to obtain lithium carbonate. The excess lithium brine, already concentrated is reinjected back into the salt aquifer [13].

Figure 10.10 represents one of the most important elements of the SQM salar process (often overlooked by analysts) is the excess of rich lithium brine (0.6–1.7% Li) after the potash production. As the diagram shows, this brine is reinjected into the porous salt and is presumably available for future brine production. This may well be the case as SQM has announced a 20% reduction of (virgin?) brine pumping. The reinjected high lithium brine, which is about five times richer than the virgin brine pumped into the potassium ponds could well replace this decrease and permit the company to realize its published increased production without affecting the salar hydrology.

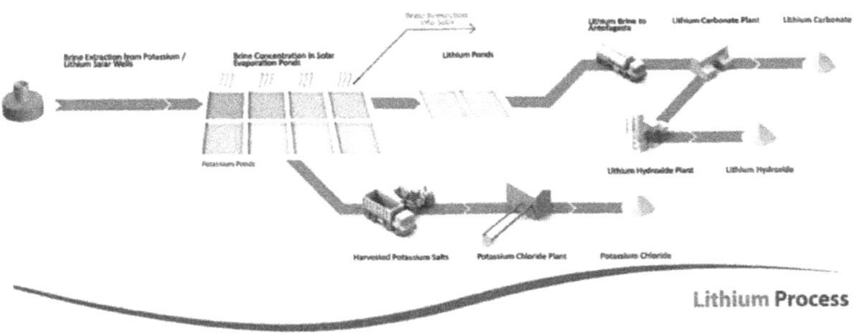

Figure 10.10 Soquimich Salar de Atacama brine flow sheet.

To satisfy the required production, SQM has drilled some 383 wells in the high-potassium—low-sulfate zone and those are feeders to the southernmost set of evaporation ponds, while the northern pond system is fed by eight high-sulfate brines (Fig. 10.11).

The red dots represent the location of the production wells. As of September 2017, 376 wells have been drilled. While SQM has the rights to pump 492 million cubic meters of brine (1.05 million MT of lithium), it voluntarily reduced the volume to 342 million cubic meters (equal to 0.82 million MT of lithium). The proven and probable reserves are reported at 1.533 million cubic meters (5.3 million MT of lithium) [14]. The potassium salts are produced at the salar while the lithium chloride brine is trucked to the plants located at the Salar del Carmen to be converted to lithium carbonate and lithium hydroxide. Soquimich reports that the current capacity of lithium carbonate stands at 48,000 MT. The company has announced that it has committed to U$321.3 million to increase lithium carbonate production to 70,000 MT in 2018 and, in 2020, announced a capacity increase to 180,000 MT in 2023. SQM also increased its lithium hydroxide capacity from 6000 MT in 2008, to 13,500 MT in 2018 and plans to expand to 30,000 MT in 2023 [14].

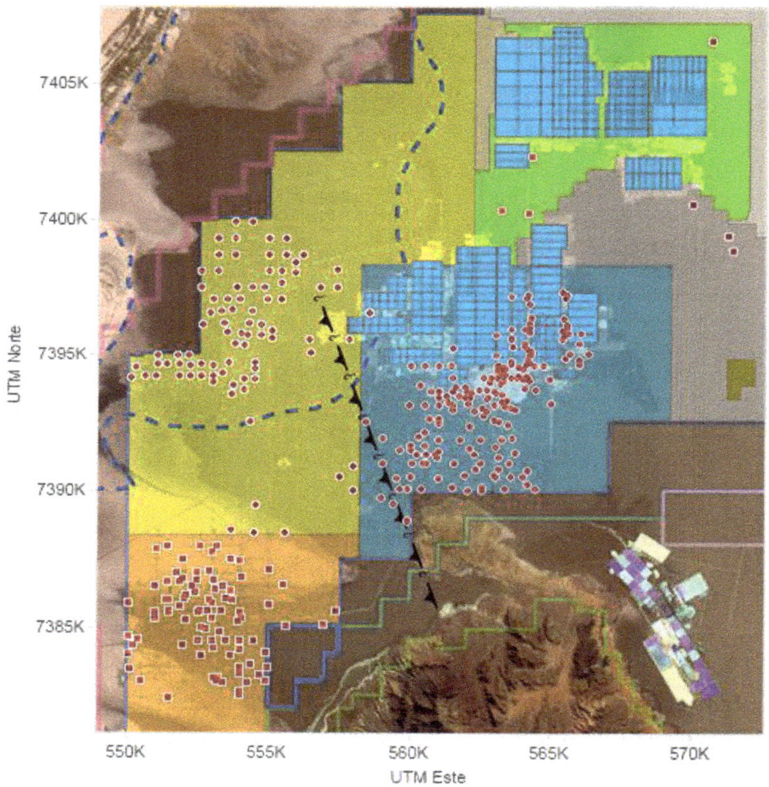

Figure 10.11 Soquimich well drilling density.

Today, SQM has developed a large operation producing potassium chloride (KCl), potassium sulfate (K_2SO_4) at the salar and lithium carbonate, lithium hydroxide, and boric acid at the chemical plant located east of Antofagasta where energy and qualified personnel are available.

To protect the SCL operations from the impact of such high brine withdrawal, a 2-km buffer zone north of the SCL claims was established. However, it is not clear whether this represents sufficient protection. (In fact, I had recommended the installation of observation wells along the buffer zone to monitor the decline in the brine levels and suggested that a formal letter be sent to Minsal—the precursor of SQM).

Recently the Chilean Government has challenged both Albemarle and SQM over the impact of brine allowed to be pumped. Environmental groups have raised the issue of the lowering of the groundwater levels at the edges of the salar, blaming the brine pumping. The published expansion plans assume that the environmental issues are resolved. As of December 31, 2020, only 10 years remain on the term of the Soquimich Lease Agreement. To date, the extracted brine represents approximately 28% of the total permitted extraction and sales limit of lithium.

10.3.4 Salar del Hombre Muerto, Livent Corporation

FMC—Lithium Corporation of America (now Livent) needed to replace the lithium chemical production in North Carolina after the low price for the lithium carbonate offered by SQM in 1997. It continued with a spodumene concentrate supply contract by the Canadian Quebec Lithium corporation. After identifying and acquiring the rights to the Salar del Hombre Muerto in Argentina, it terminated its supply contract and relied exclusively on producing lithium chemicals from the salar brines, in 1996 (Fig. 10.12).

Livent extracts lithium from naturally occurring lithium-rich brines from an area covering approximately 300 km^2. Brine containing approximately 600 ppm lithium is pumped from saltwater aquifers using extraction wells. Livent reports that the brine reserves are sufficient for 70 years of production, likely to be less with the recently announced increase of production. The brine is then diverted to an evaporation pond system, located in the southern part of the salar. Using a proprietary selective adsorption (SA) technology, the brines are first treated by selective absorption to extract the lithium, then are returned to the salar where they are concentrated by solar evaporation. This significantly reduces the time from pumping brine from the salar to processing it into lithium carbonate or lithium chloride. This reduction in processing time compares favorably to a conventional solar evaporation process, while effectively removing impurities. The resulting lithium chloride brine from the terminal pond of the system is then routed to our processing plants and processed to produce lithium carbonate at the Salar de Pocitos plant (23,000 MT) and lithium chloride (5,848 MT) and lithium hydroxide (12,411 MT) at the Guemes plant located near the town of Salta. Livent has announced plans to increase lithium hydroxide production to 30,000 MT by 2019 and then to 55,000 MT by the end of 2025 [15]. Livent has also announced plans to expand lithium carbonate production in Argentina from 15,000 MT in 2017 to at least 60,000 MT by the end of 2025 in four separate stages.

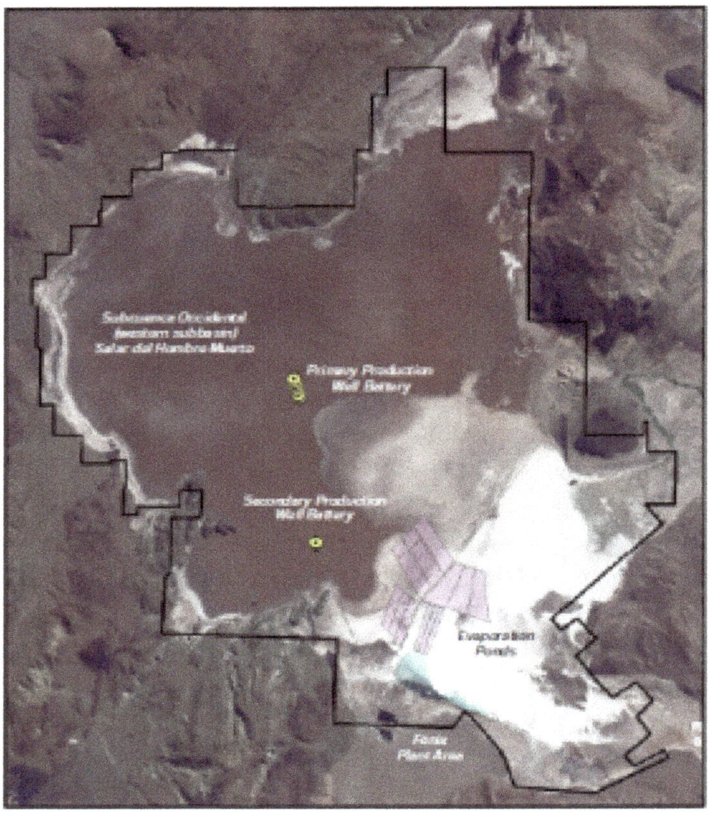

Figure 10.12 Livent Corporation Salar Operations, Salar Del Hombre *Muerto*, Argentina.

The 2018 Livent report indicates that lithium hydroxide constitutes 55% of the company revenues followed by butyllithium (26%) and lithium carbonate and chloride (6%). In 2019, Livent produced 22,500 MT of lithium carbonate and 17,500 MT of lithium chloride.

While the Foote Mineral Company focused historically on the major chemicals, lithium carbonate, lithium hydroxide, butyllithium and lithium chloride, LCA invested much in research and has one of the broadest product portfolios in the industry, providing 77 different chemical products and powering demand for green energy, modern mobility, the mobile economy, and specialized innovations, including light alloys and lubricants.

10.3.5 Oro Cobre (Olaroz–Cauchari)

The latest newcomer to the lithium supply side in Argentina is Orocobre. As mentioned, many salars in Argentina have been evaluated, some tested, but none have materialized over the last 20 years until Orocobre developed the salar de Olaroz. Orocobre is a joint venture between Orocobre Limited, Toyota Tsusho and the Jujuy Energia y Mineria Sociedad del Estado (Operations–Orocobre Limited).

Production from the Salar del Rincon has been attempted over many years, but it has not yet become an important player.

Drilling has confirmed the presence of brine grading 700 ppm lithium and potassium chloride levels of 1.3% (1.5 million metric tons of lithium carbonate and 4.4 million metric tons of potassium chloride.

While this is a brine similar to the brine of the Salar del Hombre Muerto, the low magnesium to lithium ratio of 2.2 allows the processing of brines using lime as in the Silver Peak, Nevada, operation to eliminate the magnesium.

ASX has used the contoured alluvial fan located east of the salar to construct its evaporation ponds.

The proposed product mix is 17,500 MT/annum battery-grade lithium carbonate (>99.5% Li_2CO_3) from the existing purification circuit, and 25,000 ton/annum primary grade lithium carbonate (>99.0% Li_2CO_3), of which ~10,000 MT/annum will be used as feedstock for the Naraha Lithium Hydroxide Plant, currently under construction in Japan [16].

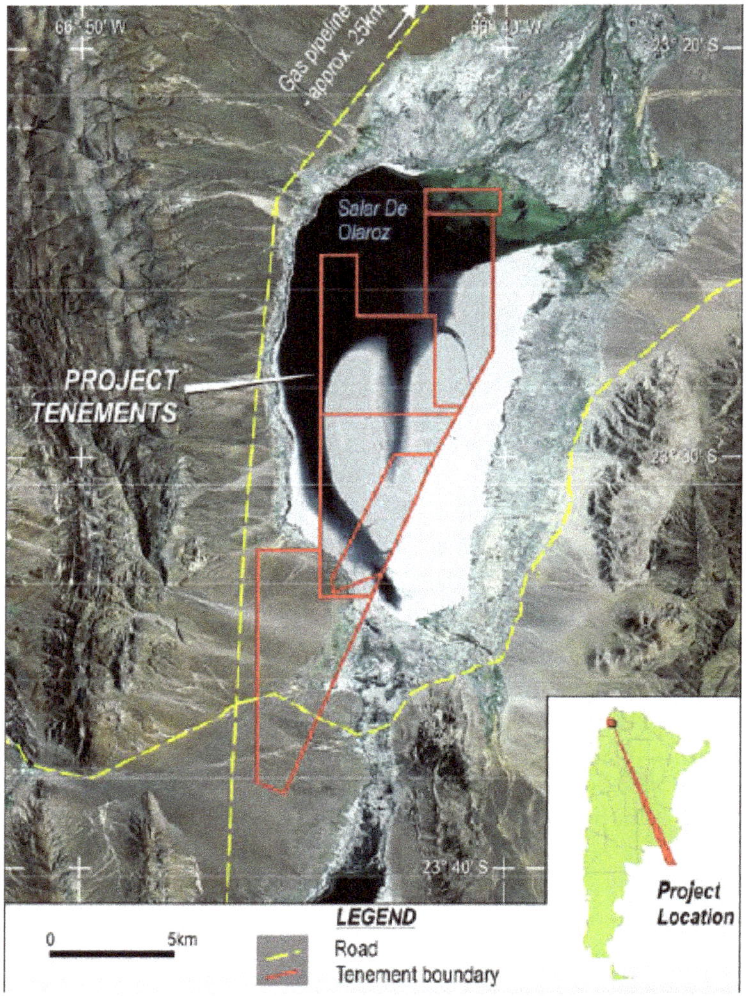

Figure 10.13 Orocobre tenements, Salar de Olaroz, Argentina.

Recently, Orocobre Ltd. announced the purchase of the smaller domestic peer Galaxy Resources for A$4 billion to create the world's fifth most valuable producer of lithium raw material for electric vehicle batteries.

Figure 10.14 Orocobre pond system, Salar de Olaroz, Argentina.

10.3.5.1 Potential lithium projects at the early stages of development

- LSC Lithium Corp (Salar de Pozuelos) changed from pilot plant construction to advanced exploration—Li resource estimate of 492,000 MT
- Galaxy Resources—Sal de Vida (East of Salar del Hombre Muerto where Livent is operating)—Lithium Resource estimate of 207,000 MT Li
- Lithium-X Energy—Salar del Diablillos—Sal de Los Angeles Project—Lithium resources estimated at 195,500 MT Li

Argentina's CAEM [17] predicts a 175,000 MT of lithium carbonate production nationally by 2025 from an estimated 38,800 MT this year. But, as Reuters points out, lithium projects often take years to ramp up. The 2025 projection remains a lofty goal.

A consortium led by South Korea's POSCO, and Samsung SDI has won a bid to build a cathode materials plant in Chile. They will jointly invest about 54 million dollars in the plant, which will produce 3,200t of cathode a year starting by 2021.

The newly formed consortium also includes the Chinese Sichuan Fulin Industrial Group and Chile's Molymet. Leadership is with

POSCO and Samsung SDI as they aim to deliver cathode materials for lithium-ion batteries for electric vehicles [18].

10.3.6 Uyuni

The Salar de Uyuni is the largest salar in the world and its resources could be enormous. Risacher has calculated a high *in situ* resource of nearly 9 million MT [19]. What is more important is the information provided by a 200-m hole, which showed elevated lithium concentrations [20].

While problems of phase chemistry are constantly raised on the Salar de Uyuni, some lithium carbonate has been produced and the phase chemistry might be ultimately resolved, possibly through novel treatment methods. Similar brines in the People's Republic of China have plagued Chinese researchers and, to date, have not been resolved and Chinese brines have not been important. Presently, the PRC relies on imports of pegmatites from Australia to feed their lithium conversion plants.

10.3.7 Talison Greenbushes: Australia (Tianqi/Albemarle)

It is interesting to reflect on how the supply of lithium has evolved. In the 1950s the primary supply of lithium was from spodumene pegmatites, and two companies controlled that production: Foote Mineral Company and Lithium Corporation of America. With the discovery of lithium brines in the Clayton Valley in the late 1960s, lithium brines became the most significant sources of lithium, first with the extremely rich Salar de Atacama brines and now recently by the overtaking of the supply by the extremely rich spodumene deposits of Greenbushes. The largest operational pegmatite is that of Greenbushes in Australia. As has been common for most pegmatites, the pegmatite was originally mined for its tin from the soft weathered upper portion. Then, the company thought to exploit the white mineral believe to be feldspar. During my visit to the property, I was shown the cores drilled on the property. Having experience with spodumene cores at the Foote Mineral Company Kings Mountain mine, it became clear that some of the intersected white mineral was not feldspar, but that the silkiness exhibited by some of the core intersections could only be spodumene, which I pointed out to the geologists. I would like to think that I had somewhat contributed to what was to become the largest spodumene operation.

The Greenbushes deposit was discovered by a government geologist, E. T. Hardman, in 1881. From this time till 1944, tin mining was sporadic and conducted with hydraulic sluices with small-scale deep mining. After the Second World War, the Tin and Strategic Mineral Company Pty Ltd started large-scale mining using modern methods [21] Due to low prices, the company ceased in 1956, and mining was taken over by Greenbushes Tin NL in 1965. Initially, this was by dredges, then from 1972 by open pit mining for tin and tantalum. Commercial spodumene production for lithium started in 1983, but the company ran into financial difficulties and was purchased by Gwalia Consolidated Ltd. in 1990, who expanded the mining operation.

The Greenbushes ore body is a highly mineralized zoned pegmatite with a strike length of more than 3 km. Its mineral reserve is unique grading 50% spodumene. This makes Greenbushes the highest-grade lithium mineral resource in the world at 2.8% Li_2O mineral reserves versus the 1.0–2.0% Li_2O from other known hard rock unzoned pegmatites (mostly around 1.5% Li_2O). Talison is the only primary lithium concentrate supplier not producing from brine operations but rather from rock mining operations. Talison's Greenbushes ore is processed via gravity, heavy media, flotation, and magnetic processes into a range of spodumene concentrates from 5.0% to 7.2% lithium oxide. Further operating modes are employed to make products that meet specific customer and application requirements. It produces two categories of lithium concentrates: technical-grade lithium concentrates—low-iron content for use in the manufacture of glass, ceramics, and heat-proof cookware; and high yielding chemical-grade lithium concentrate—used to produce lithium chemicals which form the basis for the manufacture of lithium-ion batteries for laptop computers, mobile phones, and electric cars.

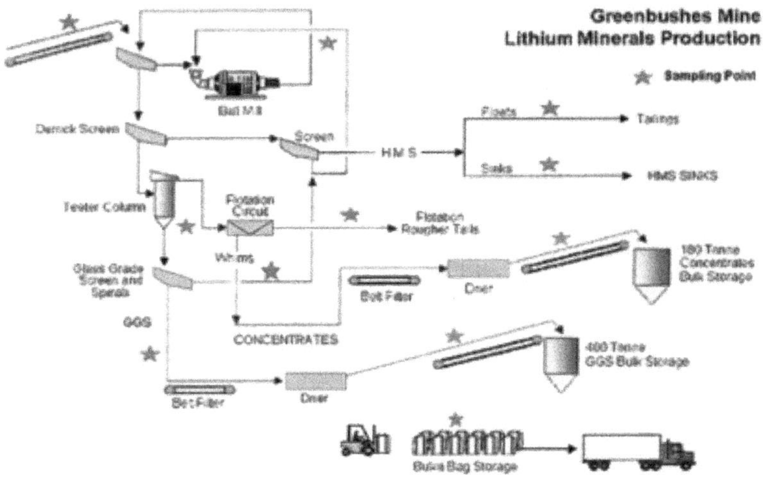

Figure 10.15 Greenbushes Spodumene operation flow sheet.

Talison is currently processing spodumene ore in two separate processing plants—technical and chemical-grade, to produce a lithium mineral concentrate of approximately 6% Li_2O. The processing plants employ gravity methods to produce lithium mineral concentrate. Talison constructed a chemical-grade to further increase the production capacity by approximately 500,000 tons/annum—from 1.34 MT/annum to approximately 180,000 tons/annum of LCE. During 2020, 88,000 MT of LCE of lithium concentrate were produced at the Greenbushes facilities. Talison currently sells the lithium concentrate only to its shareholders. In 2019, Talison plans to expand to 1.3 MT of spodumene concentrate [22]. Although Talison did not produce lithium chemical products but sold the concentrates to China for conversion into lithium chemicals and metal, Albemarle announced, in 2019, that it is commissioning a lithium hydroxide plant at the Kemerton Strategic Industrial Area in Western Australia to develop the Kemerton lithium hydroxide processing plant ("Kemerton"). Once construction is complete, Kemerton will process spodumene ore to produce lithium hydroxide product and a sodium sulfate by-product. Kemerton is expected to have an initial capacity of about 50,000 MT LCE of lithium hydroxide, with an ability to expand to 100,000 MT LCE over time (Albemarle.com).

Greenbushes lithium mine's proven and probable mineral reserves have been reported at 178.5 million tons of ore grading 2.0% Li_2O for a contained 3.6 million tons Li_2O in 2019.

10.3.8 Bikita Minerals

The Bikita pegmatite body became important in the 1950s because of the AEC program. American Potash and Chemicals processed Bikita lepidolite in Texas to separate the lithium-6 isotope to produce the hydrogen bomb. As a result of this, the lepidolite has been exhausted. Following the U.N. resolution calling for all nations to break off economic relations with Southern Rhodesia in 1965, an important source of petalite was eliminated, although a sudden "new pegmatite" was discovered at the northern border of South Africa and Zimbabwe, which was a surreptitious location for the continued export of lithium minerals from Southern Rhodesia.

Figure 10.16 Areal view of mining operation, Bikita, Zimbabwe.

In 2015, Zimbabwe produced 900 MT of lithium concentrate. By 2018, Bikita's production was forecast to total 7,500 MT LCE (roughly 71,500 MT petalite concentrate) with 50–55,000 MT of petalite exported to South Africa.

Bikita Minerals Pvt. Ltd. has an unusual variety and tonnage of commercial lithium minerals, as well as tantalum, tin, beryl, and pollucite. Its mineral reserves were estimated to be 6 million MT Li in

1961 [23], increasing to 23 million MT Li [24]. Bulk prices have been reported to be as low as \$147/MT of 4.2% $Li_2O \cdot$ petalite concentrate. Bikita Minerals has been producing lithium-containing concentrates for the ceramics industry since the 1960s. According to Garrett, the proven reserves are 23 MT of ore at 1.4% lithium oxide content, i.e., a resource in place of 150,000 MT of lithium but the USGS estimates that the production capacity is only 23,000 MT. Lithium Carbonate has never been produced from this lithium source. Matobo, Mazoe, Mutoko, Harare, Mutare, Kamativi and Hwange districts also contain some lithium ores [23].

The latest project is the Arcadia lithium project in Zimbabwe apparently the only lithium hard rock project that can produce both spodumene and petalite products. Prospect Resources expects production to reach approximately 212,000 MT of 6% spodumene, 216,000 MT of 4% petalite, and 188,000 pounds of tantalum pentoxide. This places Prospect Resources as the largest petalite producer in the world and a significant producer of spodumene [25].

10.4 Recent Lithium Producers

The tables show that the four major producers have the capacity to increase their production substantially, while the newcomer's production capacity, and their individual impact on the lithium market is limited because of the size of their resources. Additionally, it is not clear whether all will, in fact, be able to achieve the listed production levels.

Table 10.1 Active lithium operations

Producers	6% Li_2O concentrates	LCE equivalent	LCE projected	References
ALBEMARLE: Australia**		43,120		Albemarle.com
ALBEMARLE: Chile		60,000	140,000	Albemarle.com
ALBEMARLE: USA - Brine		2,200	8,000	Albemarle.com

Producers	**6% Li_2O concentrates**	**LCE equivalent**	**LCE projected**	**References**
SQM: Chile - Brine		72,000	180,000	SQM.com
LIVENT: Argentina				Livent.com
Brine		17,000	60,000	
Hydroxide			17,830	
Tianqi: Australia	1,800,000	44,880	160,000	Greenbushes.com
Wodgina: Australia	750,000		113,500	Mineralresources.com
Mt. Marion: Australia	485,000	72,000	72,000	Mineralresources.com
Pilgangoora: Australia	330,000	48,990	126,195	Pilbaraminerals.com
Mt. Cattlin: Australia	108,658	16,000	30,000	Gxy.com
China: Brines		74,000	74,000	USGS
Orocobre: Brine		17,500	42,500	Orocobre.com
Rincon: Brine		2,000	10,000	Argosyminerals.com
Manono: DRC	750,000		81,200	AVZ.com
Portugal		26,000	26,000	USGS
Zimbabwe		6,300	6,300	USGS
Subtotal: Li_2CO_3		**468,000**	**1,117,525**	
Subtotal: Li		**87,970**	**210,060**	

**Albemarle Australia production is included under Greenbushes
Note: These numbers include 35% non-battery lithium usage.

Table 10.2 Potential lithium operations

Company	Commodity	Li_2CO_3 equiv.	References
Pure Energy - USA	clays	11,500	Purenergyminerals.com
Cypress - USA	clays	27,400	cypressdevelopmentcorp.com
Thacker Pass - USA	clays	30,000	Lithiumamericas.com
Ioneer - USA	Clay; boron	25,000	Ioneer.com
Jadar - Serbia	lithium, boron	56,000	Riotionto.com
Geothermal - California	brines	16,500	Symbol.com
Geothermal - Rhine	brines	15,000	v-et.eu
E3 - Canada	brines	20,000	E3metalscorp.com
Standard Lithium - USA	brines	20,900	Standardlithium.com
Piedmont - USA	spodumene	20,000	Piedmontlithium.com
North Carolina - others	spodumene	40,000	Albemarle.com, Livent.com
Kachi Brine - Argentina	brines	25,500	Lakeresources.com.au
Searles Lake - USA	tailings	5,000	Kerr-McGee Corp
Whabouchi - Canada	spodumene	30,500	Nemaskalithium.com
Gulamina - Mali	spodumene	53,000	Firefinchlrd.com
Great Salt Lake - USA	brines	20,000	Compass.com
USMagnesium - USA	brines	10,000	usmagnesium.com
Salar de Uyuni - Bolivia	brines	20,000	Lithiummine.com
Mt. Holland - Australia	spodumene	39,000	Covalentlithium.com
Grota de Cirilo - Brazil	spodumene	66,000	sigmalithiumresoures.com

Roskill reported that global consumption of base lithium compounds in 2017, was approximately 144,000 MT LCE, representing 68% of the total consumption of lithium compounds on an LCE basis. Roskill forecasts that base lithium compound consumption is expected to grow to approximately 417,000 MT LCE by 2027, representing an 11% compounded annual growth rate (CAGR) and 47% of the total consumption of lithium compounds. Demand for lithium carbonate used in energy storage, glass and ceramics applications is expected to be the primary driver of this growth. According to Roskill, the global supply of all lithium compounds was 270,000 MT LCE in 2017. Six producers, including Livent, Albemarle Corporation ("Albemarle," previously Rockwood Holdings, Inc.), SQM, Tianqi Lithium Corporation ("Tianqi"), Jiangxi Ganfeng Lithium Co. Ltd. ("Jiangxi Ganfeng Lithium") and Orocobre Limited ("Orocobre"), accounted for approximately 86% of the total 2017 supply. Chemical concentrates are exported to China for lithium chemical conversion [26].

References

1. Arundale, J. C.,1953, Lithium in mineral facts and problems, *U. S. Bureau of Mines Bulletin* 556.
2. O'Neill, B. J., Kunasz, I, and Wright, L. A., 1969, The lithium production curve: a measure of many things, *Earth and Mineral Sciences*, **38**(6), 45–48.
3. Norton, J. J., and Schlegel, D. M., 1955, Lithium resources of North America. *USGS Bulletin* 1027.
4. Landolt, E., 1955, The lithium industry, *Journal of the Electrochemical Society*, **102**, 285C.
5. Kesler, T. L., 1942, *The Tin-Spodumene Belt of Carolinas: A Preliminary Report*, U.S. Government Printing Office: Washington, DC.
6. Connolly, J. P., O'Harra, C. C., 1929, The mineral wealth of the Black Hills, *South Dakota School of Mines, Bulletin* 16.
7. Azcarza, W., 2016, Mines Tales: Midnight Owl Mine known for lithium deposits, *Arizona Daily Star*.
8. Schreck, A. E., 1955, *Lithium: A Materials Survey*, U.S. Department of the Interior, Bureau of Mines.

9. Chagnes, A., and Swiatkowska, J., 2015, *Lithium Process Chemistry: Resources, Extraction, Batteries, and Recycling*, Elsevier.
10. Redeker, I. H., 1981, Flotation of feldspar, spodumene, quartz and mica from pegmatites in North Carolina, *Proceedings of the 13th Canadian Mineral Processors Annual Meeting, Ottawa, Canada.*
11. Evans, K., 1978, Lithium reserves and resources, *Energy*, **3**(3), 379–385.
12. Albemarle to double lithium capacity by 2025, Electrive.com, Jan. 9, 2021.
13. SQM, 2018, Production processes, www.SQM.com
14. SQM, 2020, Annual report, www.SQM.com
15. Livent Releases First Quarter 2022 Results, ir.livent.com/news-details/2022
16. www.Orocobre.com
17. www.reuters
18. Consortium led by POSCO, Samsung SDI to build $54M cathode materials plant in Chile for EVs, www.greencarcongress.com/2018
19. Risacher, F., and Fritz, B., 1991, Quaternary geochemical evolution of the salars of Uyuni and Coipasa, Central Altiplano, Bolivia, *Chemical Geology*, **90**(3–4), 211–231.
20. Fritz, Sh. C., Baker, P., Dwyer, P., Lowenstein, G., *et al.*, 2001, A 160,000 year record of Tropical Climate Variability from Salar de Uyuni, Bolivia, *AGU Fall Meeting Abstracts*.
21. Greenbushes Lithium Mine - world's largest operating mine, https://www.nsenergybusiness.com/projects/greenbushes-lithium-mine/
22. www.albemarle.com
23. Cooper, D. G., 1964, The geology of the Bikita pegmatite, in Haughton, S. H. (ed.), *The Geology of Some Ore Deposits in Southern Africa*, Vol. 2, Geological Society of South Africa: Johannesburg, p. 441–461.
24. Jaskula, B., 2010, Lithium, *USGS Mineral Commodity Summaries.*
25. Bertincourt, J. F., 2019, Petalite Characteristics and Market, Terra Studio Pvt. Ltd.
26. Roskill, 2018, Lithium: Global Industry, Markets and Outlook to 2027.
27. Jahns, R. H. and Wright L. A., 1953, Gem- and lithium-bearing pegmatites of the Pala District, San Diego County, California, United States Geological Survey.

Chapter 11

Lithium Resources and Availability

The rush to produce electric vehicles has generated a number of controversial "the sky is falling" publications that claim that there is not enough lithium to satisfy the demand for lithium battery production. The publication by Tahill holds such a position [2]. What is clear is that humans cannot predict future events. What we have observed is that everybody has gotten on the lithium bandwagon and made untenable predictions. My good professional friend Douglas Silver, a brilliant economist, stated that "weather forecasters are the most overpaid people in the world and even professional baseball players only have to hit 30% to make big bucks. The weather people get paid to look good and hope for the best [9]." I would venture that lithium forecasters are on par with weather forecasters. Thus, the analysis of lithium resources at a given time is fraught with errors.

Without going into the details of the published numbers, one should leave the determination of the lithium reserves and resources to those who have spent their professional careers investigating global lithium deposits rather than analysts who have little geologic and evaporite experience and just quote statistics from existing literature. Often these lead to significant errors as has been reported, for example, in Wikipedia. The published graph shows that around 2017, the world production of lithium was 600,000 MT [1]. The actual 2017 world production was 68,500 tonnes Li (365,000

The Lithium Legacy
Ihor Kunasz

ISBN 978-981-4968-31-7 (Hardcover), 978-1-003-37236-3 (eBook)
www.jennystanford.com

tonnes lithium carbonate equivalents). Clearly, a monumental error, which might suggest a great investment potential, but would mislead potential investors. Another important aspect of the available lithium resources and reserves is the time at which the estimates have been made. Tahill's opinion was published in 2006 [2], regardless of the fact that Tahill was not a lithium specialist (nor a lithium geologist or lithium producer).

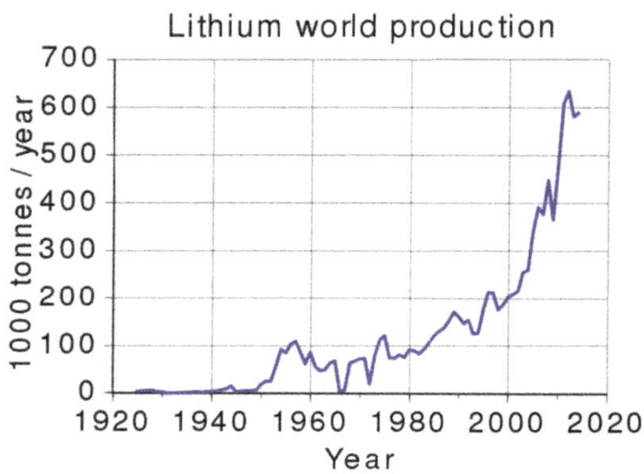

Figure 11.1 Lithium world production—erroneous.

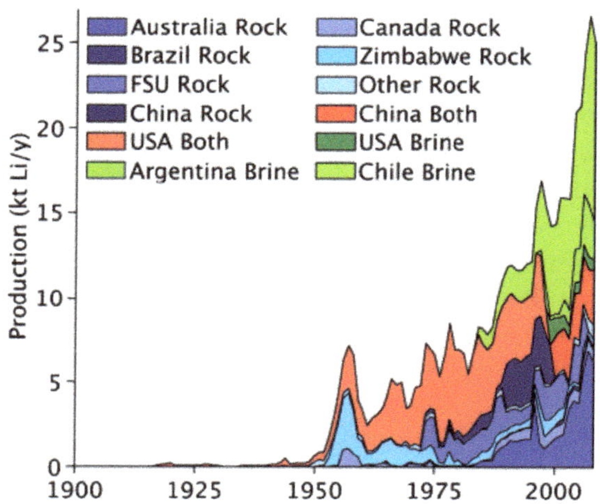

Figure 11.2 Lithium world production—actual.

Table 11.1 Lithium availability (CONAES report)

Cagtegory	United States Metric tons Li	Other Western Metric tons Li
Class A		
Pegmatites	329,100	366,400
Brines	40,500	1,290,000
Stockpiles	6100	
Subtotal A	375,700	1,656,400
Class B		
Pegmatites	47,300	456,000
Brines		
Subtotal B	47,300	456,800
Class C		
Pegmatites	2,780,900	1,969,600
Brines	77,300	3,000,000
Subtotal C	2,858,200	4,969,600
Class D		
Brines	283,700	
Cumulative Totals		
Class A	2,032,000	
Classes A+B	2,546,200	
Classes A+B+C	10,363,400	
Classes A+B+C+D	10,647,100	

A: reserves confirmed by systematic drilling

B: reserves indicated by limited exposures and/or exploration

C: resources inferred on limited exploration

D: quantities largely known but extraction

Fourteen years later, with the discovery of new deposits, his "sky is falling" warning has become outdated and irrelevant. With the sudden interest in lithium batteries for automotive and energy storage, many have become lithium resource specialists simply by re-hashing available data and publishing them as experts in the field.

Therefore, it should be a valuable exercise to review the history and evolution of resources over the past 40 years.

The most respected resource report was the one that was commissioned by the Energy Research and Development Administration. In 1975, The National Research Council Academies of Sciences and Engineering formed a National Research Committee on Nuclear and Alternative Energy Systems (CONAES), which included lithium. Thomas Kesler (author of *The Tin-Spodumene Belt of the Carolinas* and past chief geologist of Foote Mineral Company) headed the committee. James D. Vine of the United States Geological Survey; Keith Evans, from the Lithium Corporation of America and the author of the summary report; and Ihor Kunasz, chief geologist of Foote Mineral Company participated in the preparation of the report summarized by Keith Evans [3].

At that time the following was concluded:

> *. . . highly conservative assumptions have been made concerning mining recoveries and these go a considerable way to accounting for total losses. Western World reserves and resources of all classes are estimated at 10.6 million tonnes Li of which 3.5 million tonnes Li are located in the United States. Current United States capacity, virtually equivalent to Western World capacity at that time, was 4700 tonnes Li, and production in 1976 approximated 3500 tonnes Li. Production for current applications is expected to grow to approx. 10,000 tonnes in the year 2000 and 13.00 tonnes a decade later.*

However, this prediction of 13,000 MT for 2010 was clearly an underestimate as the actual world production was more than double at 28,100 MT. While that prediction was quite modest, the identified world lithium resources had increased at a faster rate to 28.4 million MT (a far cry from the knowledge available in 1976). Subsequent discoveries, particularly in brines in the southern Andes and the plateaus of western China and Tibet have increased the tonnages significantly. Geothermal brines and lithium-bearing clays add to the total. This current estimate totals 28.4 million MT Li [4]. However, many "specialists" simply canvassed the literature and came up with their own assessment of global lithium resources.

An evaluation of the following table reveals the great divergence in the reporting. Once again, I must stress that with the exception of Keith Evans, none of the "specialists" have any experience in the evaluation of lithium deposits. Evans has practical experience with the Zimbabwe and North Carolina pegmatites and the Salar de Atacama brine deposits. With the additional contributions of Tom Kesler and Ihor Kunasz, chief geologists of the Foote Mineral Company, Evans' figures are quite reliable.

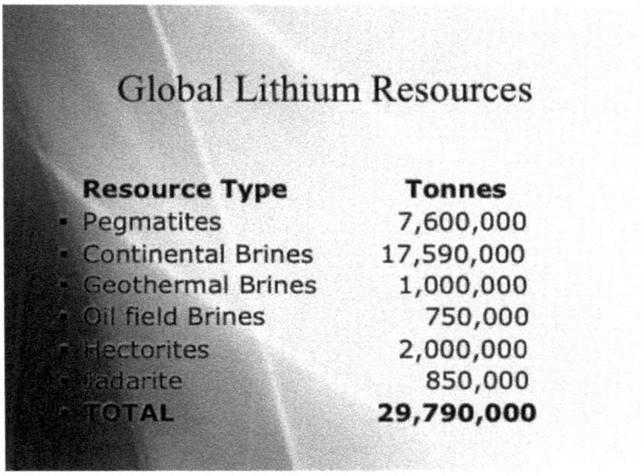

Figure 11.3 Global lithium resources.

As time went by and discoveries were made (which included jadarite—a lithium-boron deposit in Serbia [5], a new estimate published by the USGS (2020 Mineral Commodity data) increases the total global lithium resources to 72 million MT Li [6].

Table 11.2 Total lithium resources and reserves estimates

Li Resources	Reference	Li Reserves	Reference
13.8	USGS (2009a)	4.1	USGS (2009a)
19.2	Tahil (2008)	4.6	Tahil (2008)
29.9	Evans (2008b)	29.4	Yaksic and Tilton (2009)
33	USGS (2011)	13	USGS (2011)
34.5	Evans (2010)	39.4	Clarke and Harben (2009)
64.0	Yaksic and Tilton (2009)		

Note: Values are in million tonnes (Mt) of lithium (Li).

Important new lithium resources are being constantly identified (Jadar, Serbia) and existing resources (Manono-Kitotolo, DRC) resources are constantly being revised. The latest data on global lithium resources published by the USGS stand at 78 million tonnes [6].

Table 11.3 U.S.G.S. world lithium resource estimate

Country	Lithium resources tons Li
Bolivia	21,000,000
Argentina	19,300,000
Chile	9,600,000
Australia	6,400,000
China	5,100,000
DRC	3,000,000
Canada	2,900,000
Germany	2,700,000
Mexico	1,700,000
Czechia	1,300,000
Serbia	1,200,000
Peru	800,000
Mali	700,000
Zimbabwe	500,000
Brazil	470,000
Spain	300,000
Portugal	270,000
Ghana	90,000
Austria, Finland, Kazakhstan	50,000 EACH

What is not known at this time or any time in the future is the fraction of each resource that will become an economic reserve (from a production, legal and political perspective) and which will become competitive with existing large operations that count on very large reserves and resources (Australia, Chile, Argentina) and potential new production (the Democratic Republic of the Congo). These resource data are but an illustration that, with increasing interest in lithium, assuming that lithium batteries will be the only choice of the future automotive and energy storage industries (other candidates (hydrogen) are being evaluated), many more lithium deposits are likely to be identified in the future.

In the case of pegmatites, the lithium reserve/resource is simple to evaluate. The lithium minerals are fixed in the three-dimensional orebody and the calculation is straightforward (Australia, Canada, DRC, Germany, Brazil, Spain, Zimbabwe, Portugal, Spain, Austria, Finland). Conversion from spodumene concentrates to lithium chemicals and metal will entail an additional loss of lithium.

However, brines are more difficult to assess. Lithium in brines is not a fixed quantity but changes the moment pumping starts. This implies that only aquifers (those capable of delivering brine upon pumping) can be included in a resource estimate. In the case of mature salars comprising mainly sodium chloride, the resource is reasonably correct. However, in a number of immature salars, comprising mixtures of saline and detrital sediments, the reported resources by many exploration companies have included a lithium resource based on the Canadian Institute of Mining criteria (developed for metals). Since fine-grained sediments such as clays will never yield any brine, these reports must be evaluated, and the lithium contained in fine-grained sediments must be excluded. Additionally, these resources are *in situ*, not reserves. The extensive analysis by Houston points out that only about a third of the brine resource will be recovered [7]. Additional losses will be incurred in conversion to lithium chemicals. It is not clear whether or not the USGS list for Chile, Argentina, Bolivia, and the People's Republic of China includes such recovery factors.

One of the factors that are not being considered in the discussion of the long-term lithium resource availability is recycled lithium. As more and more batteries reach their performance life (6–8 years), these batteries are and will need to be recycled as exceedingly more electric cars and storage units are produced. Many companies have entered the battery recycling business. The U. S. Department of Energy reports that 11 million tonnes of lithium batteries will reach the end of their life by 2030 [8].

References

1. https://en.wikipedia.org/wiki/Lithium as an investment
2. Tahil, W., 2007, The Trouble with Lithium: Implications of Future PHEV Production for Lithium Demand, 2006, http://www.meridian-int-res.com/Projects/Lithium_Problem_2.pdf
3. Evans, K., 1978, Lithium reserves and resources, *Energy*, **3**(3), 379–385.
4. Evans, R. K., 2008, An Abundance of Lithium. https://citeseerx.ist.psu.edu/viewdoc/download?doi=10.1.1.363.1242&rep=rep1&type=pdf
5. Stanley, C. J., Jones G. C., Rumsey M. S., Blake C., *et al.*, 2007, Jadarite, $LiNaSiB_3O_7$ (OH), a new mineral species from the Jadar Basin, Serbia, *European Journal of Mineralogy*, **19**, 575–580.
6. USGS Mineral Commodity Summaries, 2020. https://pubs.usgs.gov/periodicals/mcs2020/mcs2020.pdf
7. Houston, J., Butcher, A., Ehren, P., Evans, K., and Godfrey. L., 2012, The evaluation of brine prospects and the requirement for modifications to filing standards, *Economic Geology*, **106**, 1225–1239.
8. Jacoby, M., 2019, It's time to get serious about recycling lithium batteries, *Chemical and Engineering News*, **97**, Issue 28.
9. Silver, D. 2021, Fickle fallacies in forecasting fungible commodities, Industry Newswatch, *Mining Engineering Journal*, **73**(2), 21.

Chapter 12

Lithium: Clayton Valley, Esmeralda County, Nevada

So how does one's decision lead to becoming involved in global lithium research and evaluation. It started with a thesis evaluating the F2 salt layer mined under Detroit, MI, by the International Salt Company (few people in Detroit realized that they live, quite safely, over miles of underground tunnels). It involved collecting salt samples from the numerous individual salt layers (called varves) that represent annual salt deposition. They are continuous in the mine tunnels and correlate quite well with the same salt layers under Windsor, Ontario. The purpose was to analyze the bromide-to-chloride ratios of each varve as they provide an understanding of the evolution of the depositional conditions of the Salina Salt deposited some 400 million years ago in the Ordovician Period when seas covered the Michigan and Ohio basins. The results of my study were published in 1968 [1]. Few people in Detroit realize that they live (quite safely) over miles of underground tunnels.

After my master's thesis, an unusual opportunity presented itself, which defined my professional future.

In 1916, the U. S. government geologists were assigned to evaluate the occurrence of strategic elements for the war effort. Many dry lakes in the Southwestern United States, including Clayton Valley, a depression located directly north of Death Valley, California,

The Lithium Legacy
Ihor Kunasz

ISBN 978-981-4968-31-7 (Hardcover), 978-1-003-37236-3 (eBook)
www.jennystanford.com

were explored for many critical elements including potassium. Lithium, however, was not on the list. It was only in the 1960s that the Leprechaun Mining company decided to analyze the Clayton Valley for other elements including lithium; lithium had become an important component of ceramics and greases. When Foote Mineral Company acquired the deposit and started extracting lithium from the brines, little was known of the origin of the brines and sediments in Clayton Valley. Foote's chief geologist, Bernard J. O'Neill, contacted Penn State University, asking the chairman of the Geosciences Department, Dr. Lauren Wright, if there was a graduate student who knew anything about salt deposits. Serendipity! There I was. The contact resulted in a summer job in 1967 ($825/mo. +$125 living costs) followed by a 3-year fellowship to analyze all the available data and come up with a theory on the origin of lithium in Clayton Valley. Dr. Robert Schmalz, a geochemist, and Dr. Lauren Wright, a tectonic specialist advised and guided my thesis.

Figure 12.1 International salt company mine.

So, in 1967, my wife Zen and I set out for Nevada from Cleveland and started our trip out west for my summer employment location —Silver Peak, Nevada. We arrived in Tonopah, Nevada, some fifty miles from the Silver Peak brine operation. Foote Mineral Company had rented a house for us. The house was built on old silver mining tailings (Tonopah had been famous for its silver mining history) and the house was slightly crooked. The wind whistled constantly through the windows. Zenia's sister came to visit one summer. We picked her up in Las Vegas around midnight. The temperature registered at 96°F. After she was in our tilted house for a while, she complained that the long plane ride had affected her balance because she seemed to be walking sideways! She was relieved that it was the tilted house floor and not vertigo. Since Tonopah was a transient town, she was hired on-site by a local restaurant, with the only caveat that she could not serve alcohol because of her young age.

When I arrived for the first time to report at the Foote Mineral Company Silver Peak brine operation, I introduced myself to John Bassarear, the General Manager, and gave him my travel expenses. He looked at it and laughed because we really had spent so little on food and motels. Nobody at the Foote Mineral Company headquarters thought of giving me an advance for travel expenses. I think Mr. Bassarear appreciated my arrival since there was nobody in the company who was familiar with geology. He added some money to the expense account.

The people at the Foote Mineral Company's main office in Exton, PA, provided me with very little information about the Silver Peak operation. The only inhabited place was the very small village of Silver Peak located some 50-highway miles from the nearest town, Tonopah, where we spent the summer in a small house built on the tailings from the earlier silver milling operation. Silver Peak derived its name because of the earlier silver mining in the hills on the southwest of the town (see Section 12.1).

Since several management people also lived in Tonopah, our daily routine was to drive (not exceeding 90 mph) in the company car early in the morning from Tonopah to Silver Peak and back at night. Because of the high driving speed, the tires had to be changed every 3000 miles.

Tonopah was very expensive because all the foodstuff had to be trucked in, so we drove to Bishop, CA, an hour and a half away, to do our monthly shopping at Safeway. My wife, Zenia, was starting to feel sick in the mornings and thought that it was the Tonopah water until the hairdresser told her: "Lady...no problem with Tonopah water... you are pregnant." The doctor in Bishop, California confirmed it. So, with that opportunity to work on lithium, I also benefited by becoming a father.

At the Silver Peak operation, I occupied an old building where the cores extracted from the basin were stored. I was told that originally, the building had served as a Wells Fargo station, a house of ill repute, and finally the laboratory of a starting geologist. What an illustrious history!

12.1 Clayton Valley Historical

The history of the small town of Silver Peak precedes the development of lithium by some 80 years, when in 1883, gold mining activities took place on Mineral Ridge, west of the town, and then silver further west. Consolidation of the two mining areas resulted in the name "Silver Peak Mining District" and hence the name of the town. Between 1970 and 1986, Sunshine Mining Company's Sixteen-To-One gold and silver mine produced more than 5 million ounces of silver and 300,000 troy ounces of gold [2]. Unrecorded activities suggest that the mining of silver was likely done by Mexicans.

The precious metals mining operations waxed and waned, matched by the ups and downs of the town population. In the 1865-1871 period, as many as 150 people lived there. A 10-stamp mill was treating ores and the town counted on three saloons, two stores ad three boarding houses. At other times the population was nearly zero.

Silver Peak became a more established, albeit small, town after Clyde Keegel of the Leprechaun Mining and Chemical Company acquired 15,000 acres of U. S. public domain land and ran analyses on the brine, identifying the presence of significant values of lithium and potassium. Keegel, a professional mining engineer and supervisor at Basic Magnesium, Inc., was a specialist in the field of salines and saline brines. He was also an experienced pilot and often flew over Nevada, especially over the Clayton Valley region, which lies southwest of Tonopah in Esmeralda County, Nevada. The company was dormant until 1960 when Keegel and his four associates reactivated the company for the purpose of developing saline deposits in Clayton Valley. To acquire sufficient land for development, it was necessary for the company directors to file individual claims, which they then quit-claimed to the company. This circuitous legal maneuver was necessary because of the sodium and potassium acreage limitations allowed by law. To fund the exploration of mineral deposits the company issued a public offering of Leprechaun's common stock, which realized a capitalization of $180,000. The company used this capital to begin operations under Keegel's direction and management, with its headquarters in a large mobile home at Silver Peak. Drilling and trenching continued over the next three years and the Leprechaun Mining and Chemical, Inc. proved the commercial value of the deposits, and the Bureau of Land Management (BLM) and the United States Geological Survey (USGS) converted the permits into mineral leases. Meanwhile, the company, short of funds, started to look for an established mining company to take over the leases and develop commercial production of lithium and potash for which royalties would be paid to Leprechaun. Leprechaun reached an agreement with the Foote Mineral Company in 1965, which took over all Leprechaun's leases and permits in Silver Peak [3]. An interesting aspect of the leases surfaced when Foote Mineral Company wanted to convert the leases to claims. In metals, discovery on non-contiguous claims cannot be extended to intermediate claims. Each claim must demonstrate a discovery. In the case of the brines in Clayton Valley, Foote Mineral Company argued that since lithium is contained in a liquid, discovery can be extended to non-contiguous claims. First, it was argued that a well located at the intersection of four claims satisfied the "discovery" of lithium in the four claims.. Furthermore, pumping test data proved that lithium-bearing aquifers were continuous, and so does the lithium. The government sided with Foote Mineral Company and the leases were converted to claims.

The face of Silver Peak was changed for many decades to come after Foote Mineral Company acquired the leases and started the production of lithium carbonate from the brines, the very first brine operation to ever accomplish this. (Until the discovery of lithium in the brines of Clayton Valley, lithium chemicals were produced from minerals such as lepidolite and spodumene).

There was an old, abandoned silver processing plant that, according to the Reno Gazette [4], the president of Foote Mineral Company, L. G. Bliss, president of Foote Mineral Company, purchased for $100,000 in 1964 and converted into a lithium carbonate processing plant. According to the same publication, the initial production plan was to produce 5,000,000 lbs. of lithium carbonate. However, John Bassarear, operations manager, indicated to a meeting of the American Institute of Mining that the output will be doubled the following year. Even at that time, the Gazette suggested (50 years earlier!) the feasibility of lithium storage batteries for automotive use! Today, Albemarle Corporation is the most recent owner of Foote Mineral Company (via several earlier owners, Newmont Mining Corporation, Cyprus Minerals Company, Chemetall and Rockwood Specialties). Albemarle Corporation purchased Rockwood Holdings, Inc. in 2014 [5] for US$ 6,200 million, which also included the Salar de Atacama brine operation in Chile, the hard-rock lithium mine resource, and the lithium chemical processing plant in North Carolina. At the time of the acquisition, lithium continued to be, as it was during the early days of Foote Mineral Company, the major economic driver of the town of Silver Peak. From time to time, it is augmented by several junior mining companies that hope to discover lithium brines poking around at the edges of the Albemarle property. A lithium–boron deposit has been identified on the Rhyolite Ridge west of the Clayton Valley Basin by Ioneer, and, if developed, will likely augment the population of the town.

12.2 Clayton Valley Geological Exploration

The geology of the Clayton Valley area was examined on several occasions at the turn of the 20th century [6–12]. It is only Dole (Silver Peak March in Clayton Valley) [8] and Meinzer (Big Smoky, Clayton Valley, and Alkali Flat) who analyzed brines and waters [9]. It is only the work of the author that investigated and attempted to elucidate the origin of lithium in the Clayton Valley Basin [12]. Subsequently, many other authors [13–16] have studied the Clayton Valley Basin and conducted specific analyses of the basin chemistry, but none of the studies have led to the discovery of economic lithium deposits in the Southwestern United States.

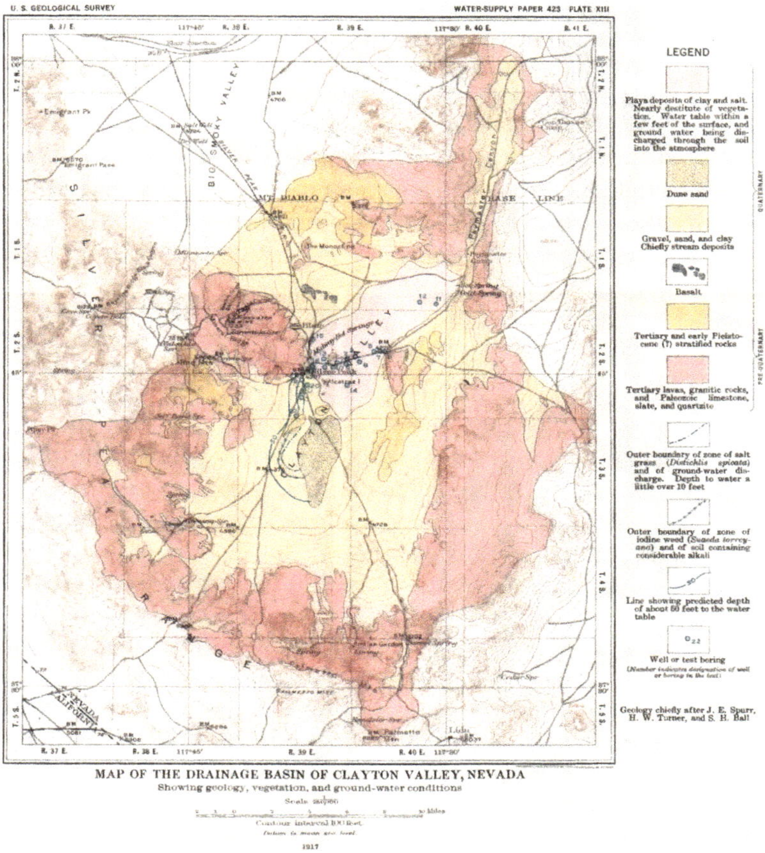

Figure 12.2 Early exploration of the Silver Peak Marsh (Dole, 1913).

In 1913, Dole conducted what was perhaps the most detailed evaluation of the brines of Clayton Valley by drilling several shallow auger wells to collect samples for analysis. This was followed in 1917 by the extensive chemical and hydrological analyses of the waters of Big Smoky Valley to the north of Clayton Valley and the Alkali Flat, east of Clayton Valley. As will be shown later, no high salinity (and therefore potassium and lithium) was encountered in these two basins. One must wonder, therefore, why the USGS decided to conduct extensive exploration drilling programs for lithium considering this available information. The same question is posed to Dajin, the company that decided to explore Alkali Flat Basin.

Dole's early drilling and sampling program is summarized in the following map of Clayton Valley (based on the geology of Spurr), the cross-sections, and the chemical analyses.

The map shows the sites drilled by Dole to document the presence of potassium, the element of interest at the time.

Dole drilled wells to a depth ranging from 11 to 40 feet. The table indicates that the brine has high salinity and that the potassium concentrations are high. The brine was intersected at depths ranging from 10 to 40 feet. After completion of the drilling, the brine levels rose to about 2 feet of the surface providing the brine samples for exploration. The potassium concentrations, while not available for all the wells drilled, show that potassium and salinities, and, by correlation, lithium, are high. The drilling and cross-sections were the first documented presence of salt layers and proved that the Clayton Valley Basin underwent significant evaporation cycles, which allowed the concentration of all the elements, including lithium.

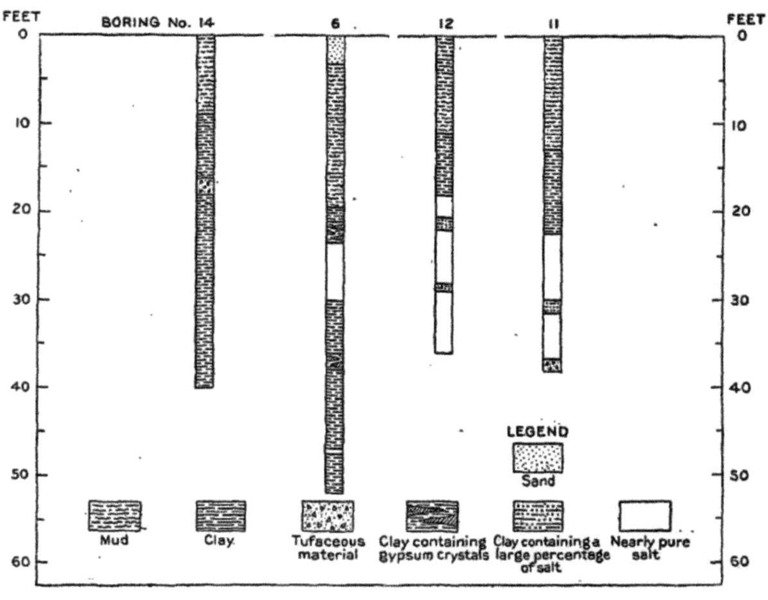

Figure 12.3 Diagram showing beds penetrated by boring numbers 6, 11, 12, and 14 in Silver Peak Marsh, Nevada.

Boring No.	Depth of sample.	Total solids at 105° C.	Potassium expressed as— Potassium (K).	Potassium chloride (KCl).	Potassium oxide (K_2O).	Percentage of potassium oxide (K_2O) in saline residue.
	Feet.					
3	15.5	33.28	0.91	1.74	1.10	3.30
6	21	33.13	.77	1.47	.93	2.80
6	40	33.75	.75	1.43	.90	2.67
11	27	32.25	.74	1.41	.89	2.76
11	35	32.05	.55	1.05	.66	2.07
12	10	26.56	.61	1.16	.74	2.78
12	20	32.90	.59	1.12	.71	2.15
12	27	32.97	.64	1.22	.77	2.34
13	16	4.15	.12	.23	.14	3.36
13	31.5	4.61	.13	.21	.16	3.43
13	40	3.38	.11	.21	.13	3.80
14	11	26.82	.66	1.26	.80	3.00
14	17	26.21	.66	1.26	.79	3.01
Average, exclusive of samples from boring No. 13		30.99	.69	1.31	.83	2.69

Figure 12.4 Potassium concentrations in borings drilled in the Silver Peak Marsh.

The low values from well (#13) drilled west of Goat Island (doubtless named for a resemblance to a similar island in San Francisco Bay) indicate that the sector is an isolated segment of the basin, without much potential.

Dole only reported the concentration of potassium. However, based on the present-day knowledge of the potassium to lithium ratio, had Dole decided to analyze the brines for lithium, he would have reported a lithium concentration of about 0.03% in his samples. As stated before, lithium was of no interest at the time of Dole's and Meinzer's exploration.

The most comprehensive geology and geochemistry related to the occurrence of lithium in Clayton Valley was prepared for a doctoral thesis by the author published in 1970. The understanding of the details of the basin geometry, chemistry, and hydrology has been developed by the geologists and hydrologists of Foote Mineral Company, but few details were published. The latest 2003 publication by Dan Zampirro, the on-site hydrologist, expanded the geological and hydrological knowledge of the basin [17].

Clayton Valley is part of what is called the Basin and Range of Southern Nevada. It is a region of structural instability formed by the stretching of the earth's crust when the Pacific oceanic plate collided with the western continental plate. The collision resulted in the extension of the continental crust in the Early Cenozoic and thus created the multiple depressions to which the "Basin and Range" terminology was given. As Major Clarence Dutton, one of the early explorers colorfully noted, the topographic map of Nevada looks like "an army of caterpillars crawling northward out of Mexico" (1886). The plate tectonic extension thus created some 50–60 basins and ranges.

The Silver Peak prospect is situated in the trans-tensional (regional shearing coupled with extension) Walker Lane structural belt, and squarely within the late Miocene to Pliocene (11 Ma to ~5 Ma) Silver Peak-Lone Mountain metamorphic core complex (SPCC), a feature that accommodated initial displacement transfer between major right-lateral strike-slip fault zones on opposite sides of the Walker Lane. The SPCC consists essentially of a ductile-deformed lower plate, or core, of Proterozoic metamorphic tectonites and tectonized Mesozoic granitoids separated by a regionally extensive, low-angle detachment fault from an upper plate of severely stretched and fractured structural slices of brittle [23].

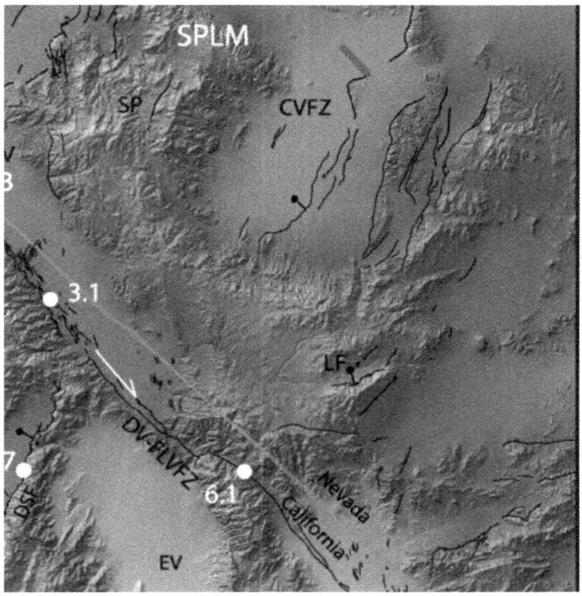

Figure 12.5 Low angle detachment fault DV-FLVFZ south of Clayton Valley, Nevada.

A discussion of the unique occurrence of lithium in Clayton Valley is important as it has all the unique conditions necessary for the creation of a lithium brine deposit.

These conditions are:

- The presence of a closed undrained structurally depression. The Clayton Valley Basin is rimmed by an arcuate zone defined by the Silver Peak Range to the west, the Palmetto Mountains to the south, and the Montezuma mountains to the east (light purple color).
- A source of lithium, either volcanic (geothermal) introduced along peripherals faults or solutions leached from earlier (tertiary) lithium-rich sediments and volcanics.
- An arid climate to concentrate the incoming solutions into a (as evidenced by salt layers).

The present extent of the Quaternary Clayton Valley playa is the latest and smallest in the history of the geological history of the basin. It is the successor of a larger Tertiary Basin as evidenced by the lithium-rich sediments Esmeralda formation exposed east of the basin. Being only about 20 square miles, the brines contained in the basin have provided economic lithium brines for over 50 years. This fact is not considered by the latest "explorationists" who simply concentrate on the white surface of a playa. More recent studies point to the presence of earlier Tertiary sediments surrounding some of the basins in North America as well as South America.

Figure 12.6 Exposure of the Tertiary Esmeralda formation, East of Clayton Valley, Nevada.

Exploration conducted by the author in the 1970s in the Southwestern United States did not identify any basins that satisfy the three conditions required for the formation of economic brine deposits. To date, the multitude of recent exploration efforts result in the discovery of any viable lithium brine deposits. Many Canadian junior companies have undertaken some intense exploration efforts in the Clayton Valley Basin picking up huge claims surrounding the Albemarle (successor to Foote Mineral Company) brine operation in the hope of identifying economic lithium brines. These newcomers do not appear to appreciate nor understand the basin hydrology and much less the implications of pumping the richer brines for over 50 years.

Around the valley, there is an exposure of more ancient sediments, which suggests that there was a much older and larger basin that occupied the structural depression during the Tertiary

Period (12 million years ago). Early studies have shown that lithium was already an important component of the earlier sediments [8, 9, 12]. My thesis work showed that the sediments from the Esmeralda formation north of the basin had lower lithium concentrations ranging from less than 20 ppm to 70 ppm Li, while those from the eastern Esmeralda formation returned values ranging from 160 to 910 ppm Li, supporting the concept that lithium solution was injected into the Tertiary Esmeralda system along with an existing fault system present on the east side of the basin. The latest exploration and drilling results from the Esmeralda formation east of Angel Island further confirm that lithium had been present during the Tertiary times (Cypress Development Corp, Clayton Valley Lithium Project Resource Estimate- NI 43-101 Technical Report, May 2018).

Figure 12.7 Montezuma Range Rhyolite—likely source of lithium to the Tertiary Esmeralda formation.

Analyses of water-soluble lithium values showed a definite localization of lithium on the eastern side of the basin. Water-soluble components from the Esmeralda exposed east of the basin returned values ranging from 14 to 115 ppm Li, while samples from Esmeralda samples exposed north of the valley varied from 6 to 16 ppm Li. Zeolites and glass contained 40 to 320 ppm Li, strongly suggesting that the sediments deposited in the eastern part of the pre-Clayton Valley Tertiary Basin were enriched in lithium and were important contributors to both the lithium-rich sediments as well as the solutions that drained into the Quaternary Clayton Valley Basin. Further evidence of the geochemical impact is provided by the lithium concentrations recorded in the volcanic ash-fall tuff of the Montezuma Range, located on the east side of the major north-northeast major Paymaster Fault structure. Samples collected across the fault zone reveal an abrupt increase of lithium to 623 ppm Li, while the values decrease substantially on either side of the fault [12]. Finally, the impact of the eastern Tertiary contribution from the Esmeralda formation is documented by the high lithium concentration in the surface sediments of Clayton Valley. When adjacent to the Eastern Tertiary Esmeralda sediments, the values in the less than 2-micron fraction (the fraction containing the lithium-bearing clay, hectorite) range from 800 ppm to 1171 ppm Li, whereas, near the northern Esmeralda formation the values decrease to 500 ppm Li.

Recent studies by Price, have suggested that lithium-rich volcanism further east in the Montezuma mountains might be the ultimate source of the lithium in the brines of Clayton Valley. Data from a Tertiary, high-silica rhyolite flow capping the western Montezuma Range and an underlying rhyolite ash-flow tuff sequence both show anomalously high amounts of Li. Lithium is most concentrated in un-weathered obsidian and in gray, relatively un-weathered ash-flow tuff, indicative of magmatic concentrations considerably higher than normal igneous rocks (Li up to 228 ppm—more than five times typical concentrations in granites and rhyolites or the median value for rhyolitic obsidians). Lithium in the ash-flow tuff is depleted from 119–192 ppm to 23–34 ppm in nearby oxidized samples, suggesting that leaching from these rocks, by weathering and other alteration, could have yielded more than enough Li to account for the resource in the Clayton Valley brines [14]. However, studies of isotope fractionation between fluid and solid indicate that the lithium in the playas of Nevada was supplied mainly through high-temperature water–rock interaction associated with local hydrothermal activity and not directly by low-temperature weathering of surface material [16].

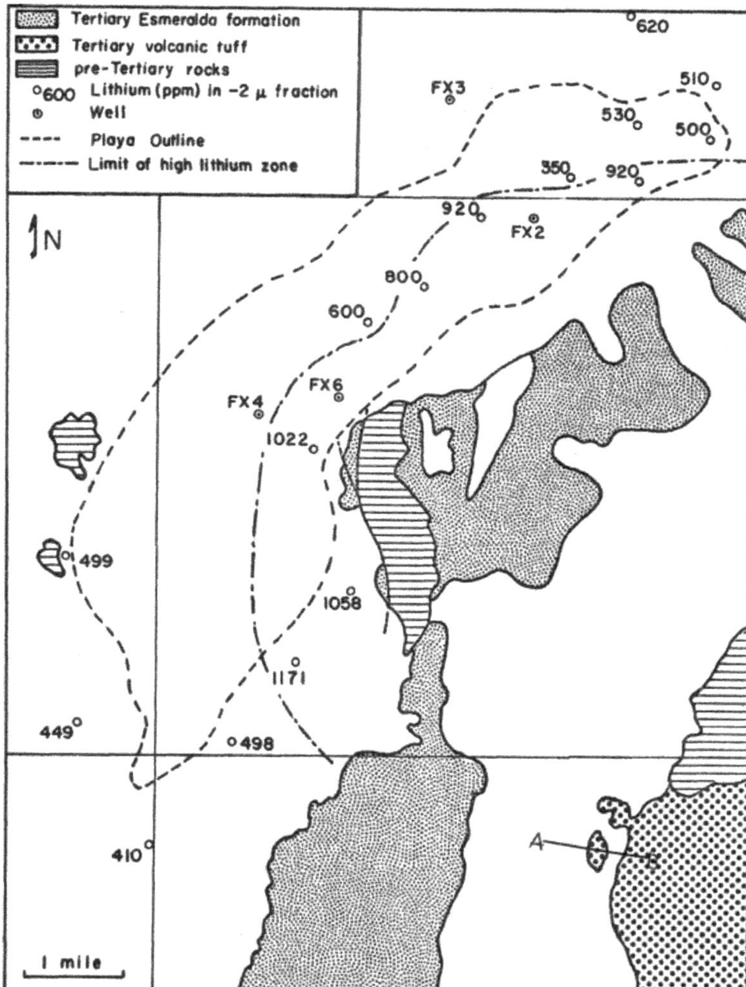

Figure 12.8 Areal distribution of lithium in the −2 μ playa samples.

Some researchers have suggested that the springs on the western side of the playa have contributed to the lithium in the basin. Although the mechanism of concentration of weak lithium-bearing solutions is clear, the source of the cations is less obvious. The hot solutions that issue from the hot springs sampled in 1968 by the author (today they are dry) that occur along the fault zones on the western margins of Clayton Valley suggest a genetic relationship. Indeed, if the hot waters are concentrated by a factor of 10 (the evaporation required for seawater to begin sodium chloride precipitation), the resulting composition approximates that of the brine.

Table 12.1 Genetic relation between spring waters and basin brine

Element	Spring (ppm)	Spring × 10	Brine (ppm)
Li	38	380	380
K	934	9,340	7,700
Na	9,220	92,200	66,200
Mg	64	640	600
Ca	443	4,430	560
Cl	13,700	137,700	95,200
SO_4	500	5,000	11,100

The chemistry of the springs may reflect a large extent by brine recirculation. White (1957) has pointed out that only about 10% of the composition of hot springs can be attributed to juvenile water and the ninety percent is derived by reaction with the country rock or mixing with meteoric water during ascension to the surface.

The presence of numerous salt beds attests to a climate arid enough to cause the initial weak solutions to concentrate into an economic lithium brine deposit. The recent suggestion [17] of dilution of the Clayton Valley brine from the northwest by leakage from the Big Smoky basin to the north is not supported as was concluded by Dole [8] who argued that the beach terraces in Big Smoky Valley are looser than the Clayton Valley Basin elevation. In fact, geothermal drilling near Blair Junction, north of Silver Peak documented geothermal brine with only 4.5 ppm lithium; these could well be the dilution of brine from the northwest [13]. However, early exploration drilling by the Lithium Corporation of America in Big Smoky Valley returned only 60 ppm Li, an unlikely source of brine to be diluted if it ever leaked into Clayton Valley.

The age of the volcanic ash-bed aquifer was determined from a sample collected from Albemarle well FM-381 at 710 feet. The ash is believed to be related to the volcanic eruption of the Bishop Tuff. The analysis by Mike Perkins of the University of Utah indicated that it directly matched the Bishop Tuff, which was dated by the U-Pb isotope dilution thermal ionization mass spectrometry at 767,000 years +/-900 years [21]. Using the age of the ash-bed aquifer and its structure, one can deduce an approximate average rate of post-ash deposition. Using the maximum depth of the sampled ash bed of 710 feet and the recorded age of 767,000 years, one can estimate that the average rate of sediment deposition into and the subsidence of the basin was approximately 1 inch per century.

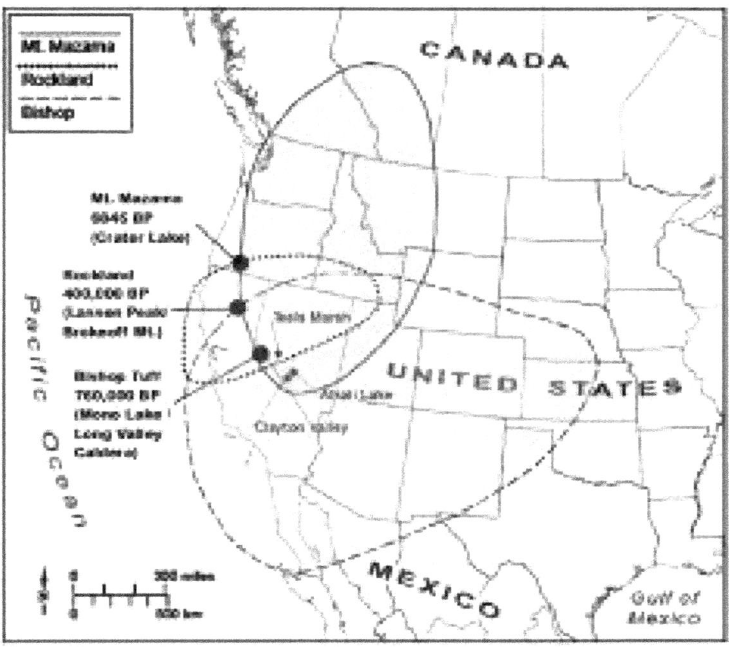

Figure 12.9 Western U.S. centers of volcanic eruptions.

Seismic data indicate that the deepest portion of the valley lies at about 4200 feet. Using the approximate ratio of total basin depth to the depth of the ash bed and applying it to the depositional time (assuming a constant rate of depression and sedimentation), one could deduce that the depression of the Clayton Valley Basin began some 4.5 million years ago.

12.3 Clayton Valley Hydrology

In the early days of the Clayton Valley operation, there was no attempt to understand the hydrology of the Clayton Valley Basin. The main question posed by the then president of Foote Mineral Company, Tony Bliss, was that, if natural brines were to replace the traditional spodumene as the main source of lithium chemical production, then what is the best commercial lithium product to be made from the brine and how can it be made [22]. Therefore, the focus was to confirm that lithium carbonate could be produced from the brine. Therefore, drilling for brine was "accidental," i.e., a site was selected and the very first production well drilled in 1966, was in the western part of the basin east of Goat Island, where the volcanic ash layer was intercepted at a depth of 250 feet and the brine was easy to capture.

The formation, which provided the brine was a permeable unconsolidated ash-fall tuff and targeted as the main aquifer for several years during the early production history. The concentration of the first wells drilled in the western part of the Basin was as high as 678 ppm [20]. Once the lithium carbonate production process was perfected, the additional wells tapped lower lithium sectors and the brine concentration dropped to 400 ppm Li [21] and kept on decreasing as the well's field expanded further east and the richer brine was slowly depleted.

Having somewhat unraveled the geological history of Clayton Valley, drilling data documented that the Clayton Valley Basin was asymmetrically tilted to the east and southeast and that the ash bed deepened toward the east and southeast. While the early production wells tapped only the ash aquifer, further exploration programs allowed not only for a better definition of the Clayton Valley stratigraphy but also identified additional aquifers. Zampirro [17], who was the well field's hydrologist, reported 5 additional lithium-producing aquifers in the eastern Clayton Valley Basin, summarized in the attached figure.

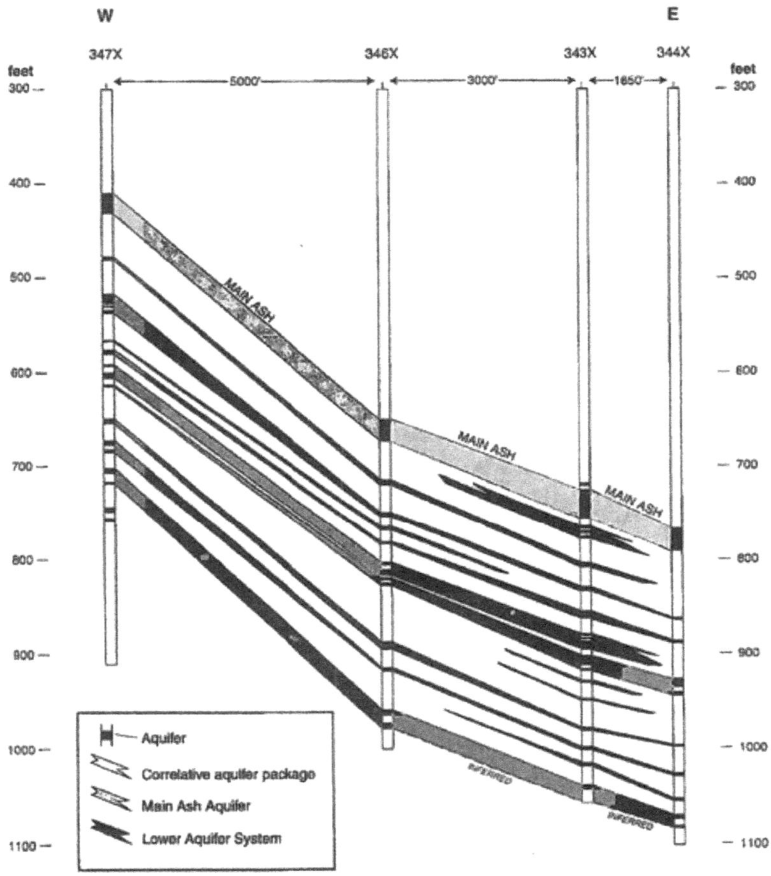

Figure 12.10 Identified brine aquifers in the Clayton Valley Basin.

These are from the top down:

- The Salt Aquifer (SAS)
- The Tufa Aquifer (TA)
- The Marginal Gravel Aquifer (MGA)
- The Main Ash Aquifer (MAA)
- The Lower Aquifer System (LAS)
- The Lower Gravel Aquifer (LGA)

The LGA occurs above the bedrock and consists of clasts of dolomite, marble, and pumice fragments. Thickening toward the north and west, its thickness ranges from 25 to 350 feet. Airlift pumping tests indicate very high transmissivity, producing up to 325 gpm but its low 40,000 TDS indicate the likely hood of quick concentration decline.

Tapping these discovered additional aquifers maintained acceptable lithium concentrations and supplied the required brine volumes that have allowed the production of lithium brine to last for over 50 years.

The LAS system consists of reworked ash, likely eroded from the peripheral areas and deposited into lacustrine depression during pluvial stages. The LAS occurs at depths of 350 feet to more than 1100 feet (I remember being on the exploration drill when we hit this horizon at about 4 am but were not sure that what we intersected was recirculation from the MAA above).

Data from the USGS show that groundwater levels have declined from 45 feet to 55 feet from 1960 to 2011 as a result of pumping. Whereas the ash layer was intercepted at a depth of about 250 feet in the western part of the basin, exploration drilling to the east promptly indicated that the main ash bed deepened to the east and decreased to depths reaching 750 feet in the northeast and was truncated by the Paymaster Fault. Further exploration indicates that the MAA ranges from 200 feet in the west and reaches over 1000 feet in the eastern sector of the playa. The available isopach maps indicate that the thickness of the ash layers ranged from 10 feet and reached its greatest thickness of 30 feet in the easternmost portion of the basin, suggesting maximum subsidence and deposition at the structural intersection of the Paymaster and Cross Faults [17]. Additional data from the southeast revealed that the southeastern portion of the eastern basin was bounded by a fault against which the brines aquifers ended abruptly. The isolation of the lithium brine was documented by a water well that had been drilled south of the Paymaster fault zone and which returned fresh water. The MAA is characterized by a reasonably high 95,000 TDS and has been the major brine source for many early years of production. Consisting of silt, sand, and gravel, the aquifer is capped by a gravel–clay aquiclude and resulted in a confined system.

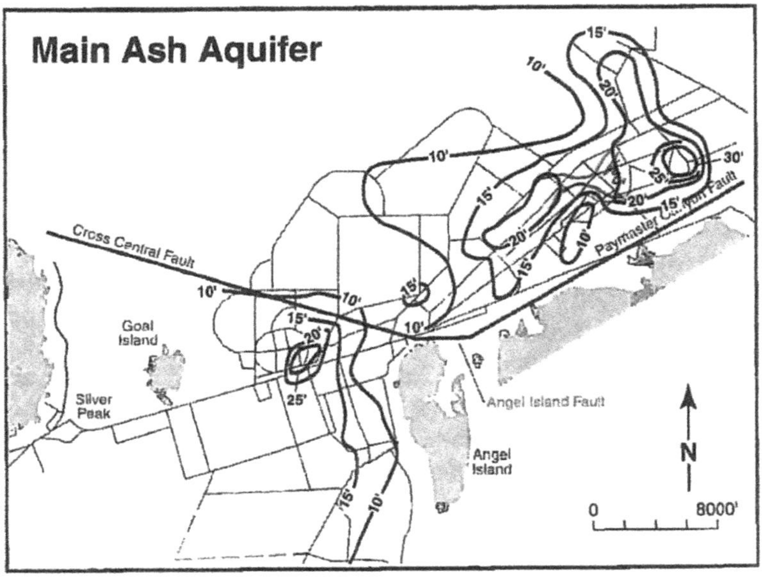

Figure 12.11 Isopach of the Main Ash Aquifer.

The MGA has provided brines that have allowed the continued brine supply to the pond system.

The richer lithium brines are slowly being replaced by peripheral weak groundwater as demonstrated by the decrease in wellfield lithium concentration. The presence of peripheral lower concentration brines has been well documented by the exploration drilling by the USGS in Ref. [18] which showed low concentrations of lithium the Foote claims, as did the over 100 Foote's exploration wells. With a constant withdrawal of brine from the basin, the lithium concentration declined from the high of 400 ppm lithium to lower concentrations, being replaced by weaker peripheral brines. Rapid removal of the brine will deplete the basin. Ultimately, the original lithium concentration will decline to the peripheral values intersected by the USGS exploration program to about 80–100 ppm lithium.

Data published by Ref. [17] for the LAS clearly illustrates, as expected, the decrease in lithium concentration to the northeast of the basin. The attempts by junior exploration companies pecking around the Albemarle claims are doomed to failure since their targets are located precisely over the dilute peripheral low lithium sectors. It is therefore difficult to understand the costly exploration programs since the drilling data will only confirm uneconomic dilute brines. So why the big rush?

The brine concentrations have declined with time due to the pumping of high lithium from the aquifers, requiring the drilling of new production wells to pump about 10–12,000,000 gallons of brine per day required to satisfy plant production of 12–15,000,000 pounds of lithium carbonate, depending on demand projections. As concentrations declined over time, pumping decreased to about 8,000,000 gallons/day and production levels have declined to about 9,000,000 pounds of lithium carbonate (USGS).

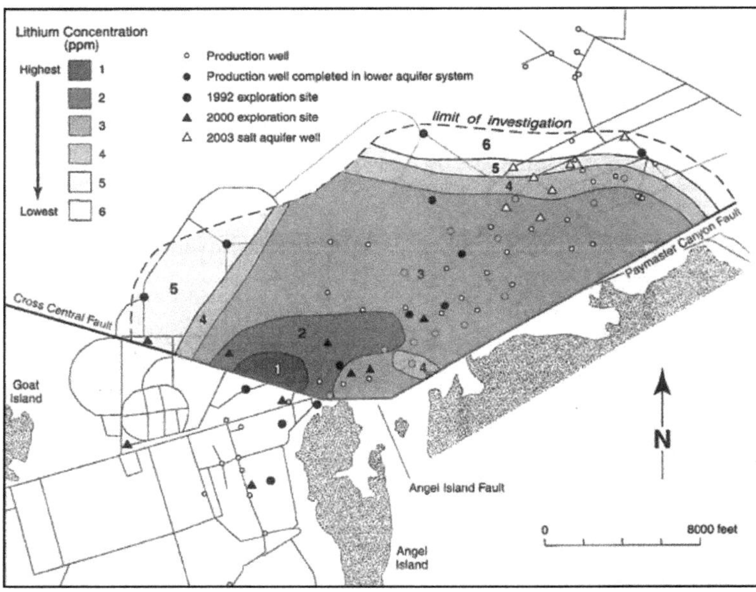

Figure 12.12 Lithium distribution in the Lower Ash Aquifer (LAS).

A model of cumulative brine removal versus lithium concentration decline, generated by the author, appears to support this inevitable conclusion. The model predicted that ultimately, the bulk of the brine would asymptotically reach the value of the peripheral less-concentrated brines. A delay occurred every time new wells with the highest lithium values came online.

While company data is considered confidential, Roskill has nevertheless printed information that supports the inevitable decrease in production from 14,000,000 to 7,500,000 lbs. of lithium carbonate [5] from 2000 to 2005 when high production levels were supported by the well field's brine concentrations.

This reduction in production is likely the result of two variables:

- The peripheral dilution of brine after some 37 years of pumping the sectors with the highest lithium brine concentrations.
- The decision to reduce production following the development of the rich brines at the Salar de Atacama.

12.3.1 Wells

In the early days of the acquisition of the leases, and before understanding the details of the stratigraphy and sedimentological makeup of the Clayton Valley Basin, which I elucidated during my thesis research at the Penn State University, there was no information on the basin or the location of the aquifers that would provide the brine for the anticipated operation. In those days, Foote Mineral Company contracted a driller to tap some brine. The first well (likely a cable tool well) was drilled on the western side of the basin, near Goat Island, and intercepted a volcanic ash aquifer at 260 feet, that provided sufficient brine used to unravel the physicochemical nature of the brine and develop a process that allowed its purification. Subsequently, as more exploration and production wells were drilled, the geometry of the basin became clearer and a model of the basin was generated which guided the future exploration targets that identified the best and richest lithium sites.

Early on, the wells were drilled with a straight mud rotary technique. While somewhat acceptable in hard-rock drilling it was quite inadequate in the type of soft sediments present in Clayton Valley, sands, muds, volcanic ash, and salt. The issue is that rotary mud drilling involves pumping mud into the drill pipe, which cools the drilling bit, and then expels under high uphole velocity the cuttings mixed with the drilling mud outside the pipe to be collected on the surface. Another issue, as illustrated in the diagram is that the injected mud coats the outside wall on its way to the surface (in red). The well needs to be "developed", i.e., flushed to dislodge the mud cake, which can reduce and hamper the production flow. The cuttings obtained by this method are difficult to identify. I remember standing next to the driller and asking him if he could tell the type of material through which he was drilling. He looks at me and said, "Man...mud is mud." So much for stratigraphic information!

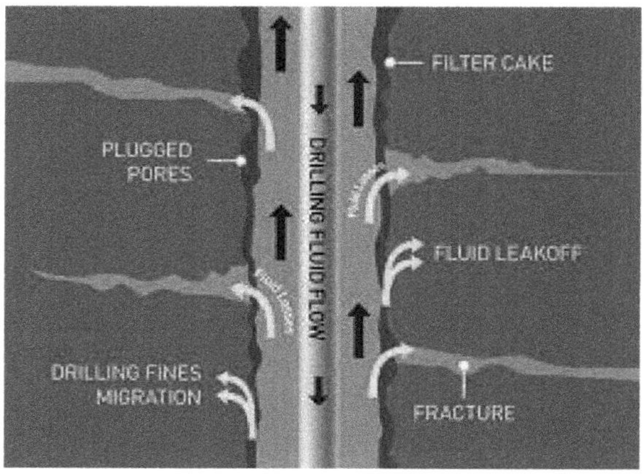

Figure 12.13 Problems with straight rotary drilling in production well completion.

This changed when two brothers from Canada, the Melhorns, invented the dual-wall reverse-circulation rotary drilling (DWRC). This technique involved a dual instead of a single tube. The drilling fluid is injected into the outer tube, and the cuttings are immediately brought to the surface for identification. The simplicity of this technique was two-fold. Once below the water table, only air and no more circulation liquid (or mud) was required, thus allowing the actual sample intercepted to be brought up and analyzed as to the type of sediment intercepted. The other advantage was that, since additional fluid was not required, one could easily determine the flow contribution from a given stratum (none from clay formations but intercepted aquifers). This technique allowed us to define the stratigraphic makeup of the Clayton Valley Basin and to identify which horizons were likely potential brine producers, in fact, Rex Bell, our staff geologist, completed an unpublished thesis in which he was able to predict the likely production volume a completed production well based on his evaluation of the data provided by the DWRC exploration wells [19].

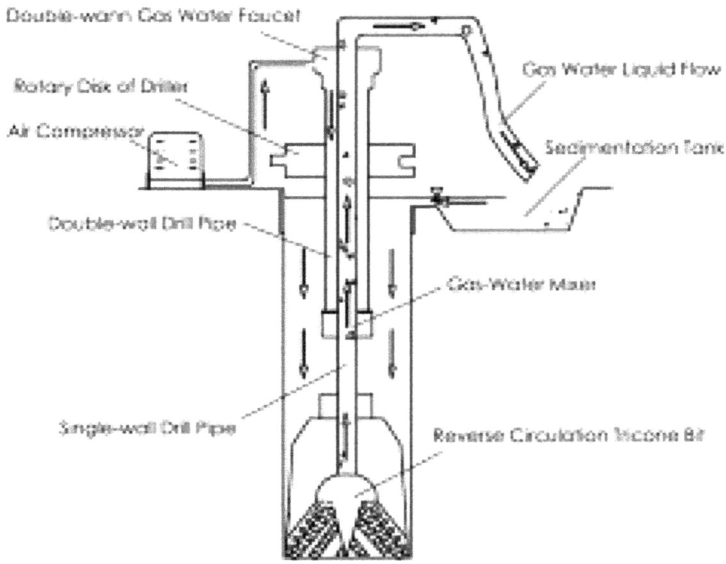

Figure 12.14 Schematic of advantage of dual-wall reverse-circulation rotary drilling.

Finally, due to the low downhole velocity of the circulating fluid in the large annular space, this technique results in the very smooth borehole wall of the production well permitting the near exact calculation of gravel pack and cement volumes required to complete the production well. This was not the case with mud rotary, because the uphole circulating fluid velocity caused substantial erosion of the well's walls. Before competing for a production well, a caliper tool was used to establish the hole geometry and calculate gravel and cement volumes. As I remember saying one time, the caliper log of a rotary mud drill hole looked like a heart attack EKG.

While some of the early wells had concentrations of 600 ppm Li, as more wells tapped formations with lower lithium concentrations the average concentration decreased to 400 ppm Li and then progressively decreased to 200 ppm Li. Since company data is not released, it is not known what the concentration is in 2020. The model developed by the author for the upper part of the basin predicted, all conditions of pumping being sustained, that the concentration would decline to the background lithium of 80 ppm Li. The drilling of additional deeper wells identified by subsequent drilling [17] and a decrease in production volumes would likely slow down the decline. In fact, the latest Albemarle 2019 presentation suggests that the well field's concentration is still around 180 ppm Li. Since company data is not released, it is not possible to evaluate such a lithium concentration.

In any brine basin, the concentrations generally decrease toward the periphery of the basin where lower concentrations exist. In the case of the Clayton Valley system, richer lithium brines tend to occur in the deeper parts of the basin where the denser brines migrated to the lower portions of the aquifers.

Figure 12.15 Collapse due to over pumping.

As the brine has been concentrated over millennia, pumping of the brines results in the replacement of the rich brines by dilute peripheral solutions. The production of the lithium carbonate at Silver Peak has progressively increased from a few million pounds to over tens of millions of lithium carbonate. The existing aquifers have been able to sustain the increase in pumped volume. However, in some cases, overstressing the well field can cause weaker peripheral solutions to short-circuit into highly permeable aquifers such as salt beds and cause the collapse of overlying sediments. If the collapse happens below an evaporation pond, all the brine contained in that pond can disappear overnight and cause an interruption in the evaporation cycle and delay the production of concentrated brines, as did, in fact, happen (Fig. 12.15).

12.3.2 Brine Ponds: The Beginnings and Early Experiments

To harness lithium from lithium brines, there are simple components of the process: wells (or trenches) to produce brine, evaporating ponds to receive and concentrate brines, and finally a chemical plant to produce lithium chemicals. However, not all brines are created equal and it is critical to understand the chemical makeup of the brines as it will control the process from the initial well feed to the final plant input. Today it seems that companies have realized this issue and proceed to solve the individually unique physicochemical character.

The initial production goal was 12 million pounds of lithium carbonate per year. This corresponds to 2.4 million pounds of lithium and requires 130 million cubic feet of 0.025% lithium natural brine.

At the Foote Mineral Company Silver Peak operation, solving the chemistry and designing a pond (ponds) system was an extensive and painful birthing process. Nobody had tackled this problem before. The president of Foote Mineral Company Tony Bliss asked the critical questions: what is the best commercial lithium product to be made from the brine? And how can it be made? Lithium carbonate was the obvious answer. However, could one recover a reasonable amount of lithium carbonate from natural solar evaporation? This question was handled by Gene Dezmelyk, head of process engineering who designed a "bench-top simulated solar evaporator," a plywood box divided into a series of four ponds of decreasing sizes of 16, 8, 5, and 3 square feet. Heating lamps provided the "sunlight." Fans and heaters "attempted" to replicate the weather (temperature, humidity, and wind velocity). The brine, when concentrated to 0.05% Li, was transferred to intermediate "ponds" until a concentration of 0.4% Li was achieved in the final pond. However, the results of the laboratory results could not be replicated in the natural environment at Silver Peak.

By 1966, the results were:

- The production of lithium carbonate was 2.4 million ponds, less than one-fourth of the goal.
- The lithium recovery was only 17% for the year against the needed 60% needed for a viable operation.
- The evaporation rate for the large 1200-acre ponds was 28 inches, two-thirds of that recorded in the wading pools.

There was a sudden realization that there was a major variable affecting the results: pond leakage. The pond design consisted of building dikes on the natural clay bottom of the Valley to retain and concentrate the brine. The later work I did on the sediments of the basin showed a substantial amount of smectite—the swelling type of clay. For a reasonable impermeable layer, the clays should have at least a 10^{-6} permeability, which the valley floor had, albeit not across the whole valley floor. Nevertheless, it was soon realized that the leakage was through the natural clay pond floor which was calculated to have leaked 5 million pounds of lithium chloride during the 1966 April to October season.

Figure 12.16 Early experimental 300-acre ponds—1965.

One of the issues that were not appreciated was the interfacial tension of the brine liquid. Pure water has a high interfacial tension; permeability testing will show acceptable permeability and leakage. Because of low interfacial tension, testing with brine resulted in a faster leakage and with lithium chloride, the leakage was even greater.

Not only was there a loss of brine through the pond bottoms but also a significant amount of lithium brine loss through the peripheral dikes enclosing the ponds. Dr. Barrett, then vice-president, traveled to Holland where Dutch engineers had designed a clay barrier. The improved pond design resulted in an improvement in lithium production of seven million pounds of lithium carbonate for 1969, 1970, and 1971. By 1973, production had reached nearly 10 million pounds from 2500 acres of ponds, the highest yet. And by 1975, the lithium operation had reached full maturity at 16 million pounds from brine filling some 4000 acres of ponds. Production costs were excellent at $1.25 per pound. Foote Mineral Company had achieved its original goal—producing the lowest cost lithium carbonate product in the world.

Figure 12.17 Present 4000-acre pond system.

The evolution of the Silver Peak lithium operation is shown in Table 12.2. When the operation achieved a steady state, some 40 wells delivered 12,000,000 gallons per day to 4000 acres of evaporating ponds. Zampirro [17] states that several hundred explorations and production wells, which ranged in depth from

70 meters to 355 meters (230 feet to 1160 feet), were drilled in the basin between 1964 and 2004.

Table 12.2 Historical lithium carbonate production

Time Period	Average Production	Average Lithium Yield
1965–1967	1.5	10
1968–1970	6.5	27
1971–1973	9.4	31
1974–1976	14.7	45
1977–1979	14.9	44

In 1975 Foote Mineral Company initiated the feasibility study on the production of 12,000,000 lbs. of lithium carbonate at the Salar de Atacama in Chile. After completion of the feasibility study and conclusion of the joint venture—the Sociedad Chilena de Litio. Ltda., production of lithium concentrates and carbonate from the Kings Mountain pegmatite deposit stopped in1986, lithium chemicals continued to be produced; the production of lithium carbonate had shifted to Silver Peak and Chile increasing a combined from 21 million to 31 million pounds in 1988. Today, Albemarle has announced an increase in capacity to10,000 metric tons of lithium carbonate at the Silver Peak Nevada Operation by 2023.

References

1. Kunasz, I. A., 1968, *Significance of Laminations in the Upper Silurian Evaporite Deposit of the Michigan Basin*, Pennsylvania State University Thesis.
2. Keith, W. J., 1988, Mineral resources of the Silver Peak Wilderness Study Area, Esmeralda County, Nevada, *USGS Bulletin* 1731.
3. University of Nevada Las Vegas Records, September 25, 2017, *Guide to the Leprechaun Mining and Chemical, Inc. Records.*
4. Reno Gazette, 28 Jan 1967, Foote Mineral Produces Lithium at Silver Peak. Output from Mineral-Rich Brines Lake to Double..., p. 36.
5. Wood Mackenzie (formerly Roskill), 2022, Silver Peak Lithium Brine: Asset Report, www.woodmac.com

6. Turner, H. W., 1900, Contribution to the geology of the Silver Peak Quadrangle, Nevada, *Geological Society of America Bulletin* 20.
7. Spurr, J. E., 1906, Ore Deposits of the Silver Peak Quadrangle, USGS professional paper No. 55.
8. Dole, R. B., 1913, Exploration of salines, in Silver Peak Marsh, *USGS Bulletin* 540.
9. Meinzer, O. E., 1917, Geology and Water Resources of Big Smoky, Clayton and Alkali Spring Valleys, Nevada, USGS Water Supply Paper 423. https://doi.org/10.3133/wsp423
10. Albers, J. P., and Stewart, J. H., 1965, Preliminary geological map of Esmeralda County, USGS Mineral Investigation Field Studies Map Mf-298.
11. Robinson, P. T., McKee, E. H., and Moiola, R. J., 1968, Cenozoic volcanism and sedimentation in the Silver Peak Region, Western Nevada, *Geological Society of America Memoir* 116.
12. Kunasz, I. A., 1970, *Geology and Geochemistry of the Lithium Deposit in Clayton Valley, Esmeralda County, Nevada*, Pennsylvania State University thesis.
13. Munk, L. A., Bradley, D., Hynek, S. A., Chamberlain, C. P., and Godfrey, L., 2011, Origin and evolution of the Li-rich brines at Clayton Valley, Nevada, USA, 11th SGA Biennial Meeting, Antofagasta, Chile.
14. Price, J. G., *et al.*, 2000, Possible volcanic source of lithium in the brine of Clayton Valley, Nevada, Geology and Ore Deposits 2000, *Geological Society of Nevada Symposium Proceedings*, 241–248.
15. Davis, J. R., Friedman, I., and Gleason, J. D., 1986, Origin of the lithium-rich brine, Clayton Valley, Nevada, *Shorter Contributions to Isotope Research, USGS Bulletin* 1622, 131–138.
16. Daisuke, A., Hodaka, K., Tetsuichi, T., and Yasushi, W., 2014, Lithium and strontium isotopic systematics in playas in Nevada, USA: constraints on the origin of lithium, *Mineralium Deposita*, **49**, 371–379.
17. Zampirro, D., 2004, Hydrogeology of Clayton Valley brine deposits, Esmeralda County, Nevada, *Nevada Bureau of Mines and Geology*, **33**, 271–280.
18. Pantea M. P., Asher-Bolinder, S., 1982, Lithologic log and lithium content of sediments drilled in Clayton Valley, Esmeralda County, Nevada, *U.S.G.S. Open-File Report*, 82–415.
19. Bell, R., 1984, Advantages and limitations of dual-wall reverse circulation rotary drilling for groundwater in unconsolidated

sediments, *Professional Engineering Degree thesis*, University of Wisconsin-Madison.

20. Garrett, D. E., 2004, *Handbook of Lithium and Natural Calcium Chloride, Their Deposits, Processing, Uses and Properties*, Elsevier Ltd.
21. Kunasz, I. A., 1974, Lithium occurrences in the brines of Clayton Valley, Esmeralda County, Nevada, In Coogan, A. H., Ed., *Proceedings of the Fourth Symposium on Salt: Northern Ohio Geological Society*, Cleveland, p. 57–65.
21. Crowley, J. L., Schiene, B., Bowring S., 2007, U-Pb dating of zircon in the Bishop Tuff at the Millenial Scale, *Geological Society of America*, **35**(12), 1123–1126.
22. Barrett, W., 1982, Internal unpublished Foote Mineral Report.
23. Oldow, J. S., Kohler, G., Donelick, R. A., 1994, late cenozoic extensional transfer in the Walker Lane Strike-Slip Belt, Nevada, *Geology*, **22**(7), 580–584.

Chapter 13

Lithium: Salar de Atacama, Chile

History does not record the thoughts of the hardy Spanish conquistador, Pedro de Valdivia, when, he, the first European to do so and crossed the forbidding wasteland in northern Chile, known as the desert of Atacama. Pedro Gutierrez de Valdivia was a lieutenant of Pizzaro the conqueror of Peru. On his way to Santiago, he stopped in the small village of San Pedro de Atacama, the most populous in the north of the Salar de Atacama, in anticipation of the arrival of the famous Spaniard Francisco de Aguirre constructed a house for a rest stop. The house still stands and has a plaque recognizing that event. The house has been converted into a tourist shop.

Pedro de Valdivia cannot have held the terrain in high esteem; more likely he prayed to his patron saint for safe deliverance. Composed of a series of dry basins, with extremely sparse vegetation and no rain to speak of in many sections of the desert, or Atacama must have awed Pedro de Valdivia 1537. In 1540, he led an expedition of 150 Spaniards and defeated a large force of indigenous natives. He founded Santiago de Chile in 1541. Governor in 1549, he began to conquer Chile south of the Bio Bio and founded Concepcion in 1550. He was captured and killed in a campaign against the ferocious local Indians, the Mapuche.

The Lithium Legacy
Ihor Kunasz

ISBN 978-981-4968-31-7 (Hardcover), 978-1-003-37236-3 (eBook)
www.jennystanford.com

Figure 13.1 House of Spanish conquistador Pedro de Aguirre—San Pedro de Atacama, Chile.

There lay in this unlikely locale, however, bounteous resources. What were the mineral riches of Chile? From the 1830s to the 1930s, fortunes were made in the mining of nitrates, an essential raw material for smokeless powder containing nitroglycerine and for fertilizers from the 1830's to the 1930's Chile was also the only source of iodine, when it was previously extracted from seaweeds and recovered from oil field gas and brines. Nitrates occur on the eastern side of the Coastal Ranges of the Desert of Atacama [1]. The caliche ore deposits are the largest ever discovered nitrate ore deposits and the only commercially exploitable nitrate deposits in the world. The caliche ore is transported to the recovery plants at Maria Elena, Nueva Victoria, and Pampa Blanca (located in Sierra Gorda).

Chilean nitrate production reached 3,000,000 tonnes in 1928. However, following the trade embargo on Germany, German scientists produced synthetic nitrates and the Chilean nitrate industry declined dramatically. By 1950 Chilean nitrates accounted for only 15% of the World Trade and had declined to 0.14%.

Today with the opportunity to produce potassium chloride from the solar data, Sociedad Química y Minera, Ltd. (SQM) has restarted the production of potassium nitrate. Up to the 1990s,

SQM had to import potassium chloride from Potash Corporation of Saskatchewan of Canada, which increased the cost of potassium nitrate production. However, when SQM acquired the northern part of the Salar de Atacama, it was able to produce low-cost potassium chloride by solar evaporation and by 2018 production of potassium nitrate increased to 1.083 million tonnes per year. In addition, SQM produces 13,000 MT of iodine [2] (lithium contributes 50% of the company's gross profit; to be discussed in greater detail further on).

In addition to its prominence in nitrates in Chile, CODELCO (Corporacion Nacional del Cobre), has been and still is a giant in copper production. Chile's copper reserves represent 28% of the world with the United States in 2nd place with 17%. However, with copper as the only commodity earning hard currency, the country was at the whim of copper prices. While today a pound of copper commends nearly $3, there was a period in the 1960s when it was about only 30 cents a pound—a financial blow after the nationalization of the three largest copper producing mines Chuquicamata, El Salvador, owned by the Anaconda Copper Company and El Teniente, owned by the Kennecott Copper Corporation. It was Anaconda, as we will discuss later, that provided Foote Mineral Company the early information on the high lithium values at the Salar de Atacama, even after Pinochet became president and liberalize the Chilean economy, the state still has retained total control of the mineral resources, although foreign investors are participating in the development of various mines.

At Chuquicamata, once a bare spot in the northern part of the desert, the largest open-pit mine in the world disgorges thousands of MT of rich copper-bearing ore. The open pit was the biggest pit in the world during the 90s, but it has lost its foremost position with the discovery of the new Escondida copper deposit which today is the largest producing mine with 1.21 million MT representing 5% of the world in 2018. Copper has been mined for centuries at Chuquicamata as was shown by the discovery in 1899 of "copper man," a mummy dated about 550 AD, which was found trapped in an ancient mine shaft by a rockfall. It is also said that Pedro de Valdivia obtained copper horseshoes from the natives when he passed through this part of northern Chile.

The modern era started when the American engineer Bradley finally developed a method of working low-grade oxidized copper ores. In 1910, he approached the lawyer and industrialist Albert C. Burrage, who sent engineers to examine Chuquicamata. It was the start of copper mining by the Chile exploration company (CHILEX) of the Guggenheim group. Their reports were positive and, in April 1911, in association with Duncan Fox S.A. and English entrepreneur Burrage started to buy up mines and claims mainly from the larger mining companies. Since he did not have the capital to develop a mine, he approached the Guggenheim Brothers. They examined the claims and estimated reserves of 690 million tonnes of extremely rich ore grading 2.58% copper. The Guggenheims also had a process for treating the low-grade ores developed by Elias Anton Cappelen Smith and organized the Chilean Exploration Company (CHILEX) in January 1912 and eventually bought out Burrage for 25 million USD ($662 million today). Smith, consulting metallurgist for the Guggenheim's sons, worked out the first process for the treatment of Chuquicamata copper oxide and directed the staff of engineers operating a pilot plant at Perth Amboy, NJ.

More recently, another natural resource of great value was discovered at the Salar de Atacama, the largest of the salt basins of the Chilean desert. Straddling the Tropic of Capricorn at an elevation of 7500 feet, the salar covers an area of about 100 square kilometers. The surface does not inspire confidence as it bristles with one-foot-high jagged peaks of crumpled salt, resulting in a very hostile environment to animals and people alike and making it totally impassable to vehicles. No doubt these and other inhospitable aspects of the land discouraged closer scrutiny nor was there any reason to suspect from the look of things that the harsh crust of discolored salt might be concealing a vast reservoir of an extraordinary liquid.

It is thus that I was thrown into what would be the longest project of my career—the lithium project in Chile's Salar de Atacama. It originated back in 1961, when Thomas Kesler, then chief geologist of Foote Mineral Company inquired about the occurrence of the lithium minerals lepidolite and petalite in Chile. On January 8, 1961, Carlos Ruiz, director of the Instituto de Investigaciones Geologicas (IIG) answered that they had been only some lithium identified by Anglo Lautaro in the nitrate deposits but none of the lithium ores mentioned by Tom Kesler had been identified in Chile.

Then sometimes in mid-60s, a geologist for the Anaconda Company, William Rudolph, was evaluating northern Chile for process water for the company's Chuquicamata copper mining operation. The cold Humboldt current and a generally easterly atmospheric circulation accentuate the rain-shadow effect of the western Andean slopes, resulting in the Atacama Desert being one of the driest areas on earth. In fact, the town of Calama, with its mining operation, and the city of Antofagasta, further west on the coast, do not have any available freshwater. Potable water is delivered from the Andes through water pipelines that tap several of the rare freshwater lakes of the Altiplano. This was the reason for Anaconda's interest in locating alternate nearby sources for its copper operation. One of the areas that Rudolph explored was the Salar de Atacama. Located some 150 kilometers east of the Chuquicamata copper operation. Rudolph did indeed find water, but it was a salinity that was as suitable for the company's purpose as seawater is for drinking. He sent Foote Mineral Company a map of the salar with the location of the brine sampling points. In addition to a lion's share of sodium chloride, the chemical analysis showed lithium grades as high as 1875 ppm lithium. By comparison, the brine at Foote Mineral Company's Silver Peak operation had a lithium concentration of 250 to 300 ppm and these lithium levels justified commercial production of lithium carbonate. By way of comparison, the Great Salt Lake of Utah brine contains 60 parts per million, the Dead Sea 20 parts per million, and the ocean a bare 0.17 parts per million. Phil Comer, vice president of Foot Mineral Company George Kneass, vice president of operation and the author met with Rudy Rudolph in Tucson to discuss the recently acquired information. Foote's interest was certainly piqued.

A more complete information on the composition of the brines of the Salar de Atacama was obtained during a 1969 exploration program conducted by the IIG and published in 1970 as an internal report [3]. Brines collected during that program, which constituted samples from a drilling campaign and shallow pits, returned lithium values greater than 1 gram/liter and potassium values greater than 10 grams/liter. Brines samples from nine holes drilled to a depth of 43 meters within the salt nucleus had lithium values ranging from 3.3 to 6.4 grams/liter and potassium values from 28.6 to 37 grams/liter.

Subsequently, in June 1970, information that the brine had a lithium content of 0.2% (2000 ppm) was published and that preliminary lithium reserves were estimated at more than 1,000,000 tonnes [3, 4].

In the 1970s, three political parties governed the country with the Christian Democrat bloc in the majority. In the elections, Allende was the candidate for the left. The largest centrist Christian Democrat block supported Allende who became the first and only president of a democratically elected communist government. Immediately Cuban and Soviet activists flooded Chile. In the end, the socialist "experiment" failed. Predictably the Chilean economy collapsed and ground to a halt. By 1972, inflation had risen to 142%, and land was privatized and given to a "proletariat" that did not know how to farm on a large scale and lacked the equipment, which it did not know how to operate. As equipment failed, spare parts from Poland did not arrive and equipment laid idle. Mining, televisions stations, and transportation were run by political appointees rewarded for their support of the regime. Stores ran out of goods and many items became available only in the black market; Enrique Arteaga, the general manager of the Sociedad Chilena de Litio joint venture (SCL), recounted how he often stood in line at 6:00 am at some store to wait and purchase some potential items; his wife would then arrive with their son at 8:00 am, switch places and he would take their son to school and she would patiently wait her turn for "something." An excellent photographic book illustrates exactly the parallel between the pre- and post-Allende situation in Chile. It was published as "Chile Ayer y Hoy" (Chile Yesterday and Today).

In October 1972, Chile suffered the first of many strikes, my good and personal and professional CORFO friend—Fernando Ide Yeager—with whom we explored the Salar de Atacama and who was a supporter of Allende told me that they were sick and tired of Allende and hoped for his fall and to be replaced by someone, preferably not Pinochet.

When Pinochet took over the government, he decided to send a few bright young men (Buchi, Ominami,) to the best economic schools in the world in the United States, Germany, and Japan where they got their economic training. Back in Chile, Pinochet told them to fix the economy and he would take care of the politics. In most communist and socialist regimes, politics and economics are intertwined and controlled by a centralized government, which eventually fails. My wife and I lived in Santiago from 1988 to 1990, we witnessed Pinochet's economic "miracle." We hired a maid who carried a Social Security booklet into which we paid the required sum of money; even she, in turn, was able to hire her maid to come and help with her house chores. While Pinochet was maligned, he made Chile an economic miracle. In the late 80s, there were already giant supermarkets in Santiago, such as "Jumbo," which sold everything from food to tires, and had 24 cashier lanes; they even had specialized parking spaces reserved "para la futura mama"—for the future mother. For an American couple, life was good, elegant clothing stores lined the various avenidas; the only thing that was not available then was furniture; you had it made by very talented woodworkers but there was no available cloth for chair covering which we had to bring from the United States. Ironically, the Soviet Union, which failed in forcing its system into Chile and besmirched Pinochet, asked him to be interviewed to see if his policies might be workable for the Soviet Union!!!

Up to the end of the Allende regime in September 1973, Chile was basically a one-commodity country—copper; when prices were high, the economy moved along; under a low-price scenario, the economy suffered. After Pinochet, a true economic revolution took place with the development of several industries. Chile realized that their kiwi ripened earlier than the New Zealand fruit one and penetrated that market; they developed a wood industry with shipments to Asia. Today one eats Chilean flash-frozen fish delivered overnight to US and European restaurants. Chile grows and exports excellent grapes, blueberries, cherries, and apples. In the 1970s and 1980s, I was able to purchase several outstanding red wines for less than 1 USD. Then, there were only a few red wines: Cousino Macul Cabernet, Pinot Undurraga, Santa Rita Reserva Familia, and Vina Tarapaca. Soon thereafter, foreign winemakers invested in the Chilean wine industry modernizing it (Mondavi, Grand Marnier L'Apostolle, and others). In 1980, with the introduction of new technology, wine production increased, and the number of wineries grew from 12 in 1995 to over 70 in 2005. Chile is now the 7th largest producer of wines in the world (12 million liters in 2019) and the 5th larger exporter. Chile has remained free of the phylloxera louse which means that the country's grape vines does not need to be grafted with phylloxera-resistant rootstock. Chile maintains the largest planting of Carmenere in the world; the grape first gained popularity in the

South American country when cuttings were brought from Bordeaux around 1850. At the time, many of the cuttings were confused with Merlot. In fact, Chilean growers called the grape "Merlot Peumo" after the Peumo valley. DNA testing in 1994 proved that the grape was indeed Carmenere not Merlot.

At that time, Foote Mineral Company was deeply involved in its technically successful Silver Peak brine operation. Lithium carbonate from that source and its spodumene mining operation in North Carolina adequately supplied the 1970 post-AEC program market. The Chilean lithium opportunity was not pursued; however, such high lithium brine, if true, could not be ignored. On July 8, 1970, Bernard J. O'Neill, the successor of Thomas Kesler, Foote Mineral Company chief geologist sent a letter to Carlos Ruiz, executive director of the Geological Institute, referring to June 22, 1970, *Christian Science Monitor*, and June 26, 1970, *Mining Journal of London*, press releases regarding the Salar de Atacama. He requested any information on the exploration conducted by the Institute; Carlos Ruiz answered that such a report was confidential but that the concentrations indeed exceeded 1 gram per liter of lithium and 10 grams per liter of potassium and that samples from drill holes showed values up to 6 grams per liter lithium and 35 grams per liter potassium. While receptive to Foote's inquiry, Ruiz requested some information on the process used at the Foote Mineral Company Silver Peak brine operation. Since Foote had solved several technical problems related to the production of lithium carbonate from a complex brine and was the only such company to successfully produce lithium from brine, it was reluctant to divulge any information on the process. Communications, however, were re-established in 1974 after the fall of the communist government of Allende in September 1973. Suddenly the lithium project at the Salar de Atacama became the highlight of my career. I contacted Dr. Corvalan, then director then of the Institute of Geological Investigations (IIG) a branch of CORFO and we received an invitation to visit the famous salar in May 1974. As an aside, Bulletin 29 of IIG mentions that Dr. Wayne Barrett, vice president of Foote Mineral Company, and the author visited the salar on May 1, 2, and 3. In those days, flying into Santiago was an interesting experience. Since the present airport, Arturo Merino, was not yet built, flights would land at the old airport of Los Cerillos around 6:00 am. Because the Pinochet government had imposed a curfew until 8:00 am, the Lufthansa crew served us breakfast and at 8:00 am we were allowed to deplane. Since there was a rigorous currency control, we had to document all in-country expenditures upon departure. It was odd to see how many money exchange places sprang up in Santiago. You would be accosted in the street and asked, "dolares... dolares...;" if you agreed you will be taken to an upstairs exchange office where you thought you would get a better rate; in fact, these were all government exchange offices and the government knew precisely what the flow of currency was in the country.

We flew to Calama via Antofagasta on LADECO airlines (Linea Aera de Cobre), an airline specifically created to fly the management of the enormous Chuquicamata copper operation, located north of the town of Calama. Today LADECO has been absorbed by LAN Chile (the national Chilean airline). In Calama, we were met by the Institute's geologist Tomas Vila and drove 90 kilometers east to the small village of San Pedro de Atacama, located north of the salar; we stayed at the "Hosteleria," a quaint little posada. In San Pedro de Atacama, we had the opportunity to visit with father Le Paige, a Belgian Jesuit, who had built a wonderful museum, which holds some of the oldest known mummies in Chile. A mining engineer by training, he was able to date the various mummies and artifacts accumulated over the centuries. Dr. Barrett and I were treated to a tour of his museum, and he invited us to sit in two lounge chairs facing an opening in the eastern wall to witness a most gorgeous sunset over the Licancabur volcano rising over 15,000 feet on the border of Chile and Bolivia. He was an amazing historian who spent his life unraveling the history of the Atacama Desert. On November 10, he sent me a thank-you note for the two colored photographs that he hoped to publish someday. This wonderful Jesuit passed away in 1980.

At the Hosteleria, we discussed the salar project with Tomas Vila and he showed us a topographic map of the salar describing what exploration the IIG had done. We found out that the project was in fact assigned to Dr. Guillermo Chong, a paleontologist by training and a professor at the University of Antofagasta. He was the district director who had undertaken a six-month study of the salar and co-published results in the above-mentioned bulletin [3]. Dr.

Chong invited me later to present a lecture on lithium brines at the University of Antofagasta.

From San Pedro de Atacama, we drove south via a decent graded dirt road built on the east side of the salar. The road services the small villages of Toconao, 30 kilometers to the south, Peine a further 60 kilometers, and Tilomonte a further 12 kilometers. From Tilomonte, the road turns West to Tilopozo, the location of the spring, and then follows the base of the Chepica Peninsula which divides the southern part of the salar into two subsectors. It took us 3 hours to reach the small CORFO field operation camp. This travel routine became common during the early days of our activities at the beginning of the project. On the way to the salar, we were stopped at 5 military checkpoints and searched for possible explosives or firearms. One must remember that our trip took place just eight months after the end of the Allende regime and the new government was wary of terrorists in this area bordering Argentina. In fact, later, we could see some explosions on the slopes of the Andes, probably some sheep stepping on mines.

We found out that the CORFO exploration work consisted of 50 shallow pits excavated along a rudimentary 20 kilometers road built on top of crushed salt. At the time of our visit, they had drilled a 48-meter core hole near the camp indicating a thick sequence of porous salt. In fact, later drilling showed that the salt in the center is greater than 200 meters thick. CORFO had blasted shallow pits on one-half kilometer centers and collected and analyzed samples of brine that was 1 foot below the salt. Dr. Barrett and I collected six samples at pits 8, 24, 37, 40, 44, and 49 to be cross-checked at the Foote Mineral Company laboratories in Exton PA. We returned with our samples to Calama to catch a noisy DC-4 return flight to Santiago. The interesting part was to clear customs in the United States. In 1975 plastic bottles were not exactly leak-proof and during the flight, our samples depressurized and leaked. At our arrival at JFK, the customs agent checking Dr. Barrett's luggage noticed a yellowish liquid; the yellowish salar that had leaked into Dr. Barrett's underwear, which he had used to protect the plastic bottles from breaking. It did cause quite a disgusting reaction from the inspector who promptly cleared us.

Figure 13.2 Original 20-km Salar de Atacama road constructed by CORFO (Corporacion de Fomento).

Although there were some discrepancies between the Chilean and the Foote chemical analyses, the samples proved that the salar brines were indeed very rich in lithium.

Foote Mineral Company promptly proceeded to organize a trip to Santiago to meet with Chilean authorities and start drafting a feasibility study to undertake a complete evaluation of the chemistry of the brines and evaluate a process to produce lithium carbonate. Philip Walsh, Dr. Barrett, Fred Luckenbach, and Enrique Puga, our Chilean lawyer, made up the negotiating team. However, reaching an agreement was not a walk in the park. Negotiations had become bogged down due to the ever-changing Chilean bureaucracy. Dr. Barrett recalls that he, Enrique Puga, Phil Walsh, and Bob Koenig, members of the Foote Mineral Company board of directors and well-respected mining executives in Chile, were invited to meet with general Pinochet in his Viña Del Mar residence. After about half an hour of Spanish conversation, President Pinochet suggested to an associate something like "why don't you get off your duffs and get the project moving." The negotiations resumed promptly. During that time Dr. Barrett had the opportunity to talk to a group of scientists from the Nuclear Energy Commission. It was a summary of the scientific paper that Dr. Barrett and I had published [5]. Barrett argued that based on my lithium reserve estimations, the salar had a fusion energy potential larger than all the oil reserves in the Middle East.

Table 13.1 Comparative chemical analyses by IIG and Foote on samples collected by Foote Mineral Company

	Brines												
	Mg		Na		K		Li		SO_4		Cl		Ca
Plot No.	**IIG**	**Foote**	**IIG**	**Foote**	**IIG**	**Foote**	**IIG**	**Foote**	**IIG**	**Foote**	**IIG**	**Foote**	**Foote**
50	1.31		9.9		2.5		.205		1.64		20.04		
49		.95		7.65		2.11		.240				16.34	.285
45	1.31		9.6		2.6		.21		1.95		19.9		
44		.98		7.03		2.21		.170				16.42	.230
40	1.01	.81	10.5	8.13	1.98	1.80	.16	.104	1.99		19.6	16.11	.0227
37		.94		7.33		2.05		.149				16.34	.0256
32	1.12		10.0		2.26		.17		1.54		19.9		
24	1.46	1.09	9.1	7.39	2.54	2.18	.204	.152	1.68		20.2	16.35	.0273
16	1.37		8.44		2.44		.299		1.49		20.0		
8	1.07	0.79	10.28	8.57	1.86	1.66	.158	.090	1.83		19.6	16.04	.0252
Average	1.235	.93	9.83	7.68	2.31	1.56	.187	.134	1.73		19.9	16.27	0.254
pH about	7.10												

Once the feasibility agreement between the Corporation de Fomento, the Chilean Development Agency, and Foote Mineral Company was signed in January 1975, work on the salar started. Two of us: David Coghlan, director of Chemical Engineering, and I started to travel regularly to the salar to set up the necessary activities to evaluate the feasibility of producing lithium carbonate from the rich but chemically complex salar brine.

In 1975, the old Los Cerillos airport in Santiago was still drab and had practically no facilities. In Santiago, however, we stayed at the Crillon, a wonderful and colorful hotel right near the Plaza de Armas, where the overthrow of the Allende regime took place. We could still see the bullet holes in the Palacio de la Moneda. As a first-time visitor to Chile, one was exposed to the wonderful taste of pisco, something unknown elsewhere in the world except in Peru. A distillate from the grapes of the Elqui Valley, a 90-proof alcohol is produced and used to make pisco sour, the Chilean version of our whiskey sour. Pisco tastes a little bit like grappa, and it is an acquired. it was interesting to discover that, in those days, "whiskey" in Chile meant exclusively scotch and it was almost a ritual to bring a bottle of Chivas Regal as a gift. Later, I was able to convince my Chilean friend to try bourbon which elicited a terrible reaction at first but, with time, gained warm acceptance.

David Coghlan and I traveled to the salar to acquire information on weather data, confirm evaporation rates, locate and test suitable clays for evaporation ponds, and drill test production wells for brine delivery to the ponds. We also collected several barrels of brine that we shipped to Foote laboratories in Exton, Pennsylvania for process testing. At the Calama airport, Dave and I were met by CORFO personnel and driven to San Pedro de Atacama, a sleepy village east of Calama and nested at the foothills of the Andes, where we spent the night at the Hosteleria, the same little hotel where Dr. Barrett and I stayed during our visit. The next day, we drove to the CORFO camp located in the southern portion of the Salar de Atacama. Some 60 km south we passed the village of Toconao, where weavers were offering their colorful rugs. Unfortunately, with the increased tourist development of the region, fewer weavers create such wonderful rugs. In those days, travel was quite difficult since there were no real roads, but only bumpy sandy trails bordering the salar. In fact, the same description and character will apply to most the salars of the world. all the while admiring the magnificent Andean Mountain chain rising to 18,000 feet with its snow-capped peaks and smoking volcanoes, a wonderful landscape that will be our companion for many years to come.

Finally, after some three hours of driving, we arrived at the campsite. CORFO had constructed a 20-kilometer northeast road onto the salar surface, the only way to access the brine sampling points. The crust of the salar is one of the most rugged physical structures ever created in a desert. Its surface covers 900 square kilometers and is completely impassable. Over the millennia, as brine evaporated, it formed polygonal cracks, which jut a foot or more; wind and sandblasting result in razor-sharp points, which tear up boot leather in a very short time, making walking over the surface of the salar extremely difficult.

CORFO, our partner, built a rudimentary field camp below the village of Peine, located east of the salar. In those early days, food was always an issue—a list would be posted daily and read as follows: Monday-no meat, Tuesday-eggs, Wednesday ...., and so on. We complemented our meals by eating in the village. Later, Foote Mineral Company constructed a better camp with a dining room, a kitchen, and acceptable beds and showers. I was exposed for the first time to "calephons"—an ingenious radiator-type gas water heater that provided warm water on demand. These very efficient systems seem to gain acceptance in the United States.

Although travel conditions to and from the camp and the southern portion of the salar were not treacherous, they were problematic. Many times, we got stuck in the soft sands or slipped on the muddy sections of the road near the salar after torrential rains during the infrequent "Invierno Boliviano"—the "Bolivian Winter" as it was called (Such an event was mentioned by the geologists of the Instituto de Investigaciones Geologicas in 1969). It is caused by an anticyclonic system, which comes from Bolivia and discharges torrential amounts of rain that batter the salar in February-March.

One night, we were awakened by the villagers who asked us to come to a usually dry creek; that night it was full to the rim and the water roared down with an incredible sound. It tore the small plots of corn that the village grew before it would dissipate into the salar. Such torrents pick up silt, sand, and clays and become so viscous that they can lift boulders as big as cars and deposit them several miles from their original location. Later, the ubiquitous desert wind blows away the fine sediments leaving isolated boulders in the middle of nowhere and people wondering how they ever got there.

Figure 13.3 Shallow exploration drilling (the author, and assistant, Rex Bell).

The people of the desert of northern Chile, as in many deserts, always settled near a source of water, much as Mormons did in Utah. In this instance, Peine—a village of some 150 people settled near a spring. In the small drainage valley of the spring, the villagers grew corn. The amount of water was controlled by a village committee, which assigned the amount of water allotted to each family. It appeared that the village was controlled by three families: the Chailes, the Morales, and a third family whose name escapes me. Since they were isolated, they produced some electricity using diesel. The electricity was on for about two hours every evening. I remember playing ping-pong with the local boys. One evening, as I was ready to return the ball, everything went dark and that was the end of the electricity for the night. They had run out of the allotted amount of diesel.

We used the local labor force in pond construction, brine sample collection, and other various jobs. As they were paid, certain members of the village were able to improve their lives. Some bought a used pick-up, others added to their homes. The electricity issue was partially resolved as they used 12-volt adapters which they hooked up to their pick-up batteries and powered their televisions and other appliances. Since they had to show up for work the next day, they would park their pick-ups facing downhill, which they would roll in neutral and then kick-start the car. The battery would be recharged during the day on the job and continued their battery routine in the evening. Foote Mineral Company build a camp and funded the school, where Maria-Inez, the wife of Felipe Anguita, the salar operations manager, devoted much time. Anguita also improved the agricultural tradition of the village by building a hothouse, which permitted the harvesting of crops twice a year instead of the century-old single annual harvest.

An interesting observation was the possibility of inbreeding. I recall when returning from the salar, we came upon a bicyclist who did not hear us even though our truck was practically on top of him. I also noted this when I would speak to the villagers. My Spanish is quite excellent, having learned it "a la Chilena;" yet I would have to repeat myself.

Figure 13.4 Kindergarten school in Peine funded by the SCL Joint-Venture (wife of the Regional Superintendent and teacher, Maria-Inez, and wife of the Salar Manager, Felipe Anguita).

Figure 13.5 Helicopter-supported exploration drilling.

I worked with mining engineers, Rigoberto Godoy and Fernando Ide, in designing an exploration program to test the quality of the brine over the total surface. Crossing and sampling by walking over 30 kilometers on 5-kilometer sampling centers was thus inconceivable. The decision was made to use a helicopter. My staff geologist, Rex Bell, became much involved in this and later projects traveling to Chile and overseeing drilling and conducting pump tests in the completed wells to predict production volumes. The concept was to prepare 6-feet, 4×4-inch, square posts and drop them on the salar on a 5-kilometer grid pattern. Rex and I encountered our first problem. Having not worked on pure salt salars, we assumed the salt to be white, therefore we painted the posts brown. However, we found out that the Salar de Atacama salt surface is brown because of the millennia of sandstorms, which regularly occur and fill every nook and cranny of your body. That we could not locate one of our posts as they blended perfectly with the brown salt. We decided to time our flight to coincide approximately with the 5-kilometer sampling points, where to helicopter would land and we would take our portable drill equipment, the 20-gallon container, and the freshwater pressurized tank. Since the brine was found at 1 foot below the surface, we would drill to a minimum of 3 feet, often deeper. Since the drilling pulverized the salt causing clogging, we would inject a minimum about of freshwater to dissolve the salt and facilitate deeper drilling. Upon completion of the well, we pumped out several volumes of the diluted brine into the 20-gallon container before collecting the actual brine sample for chemical analysis. Once the sample was collected, the contaminated liquid was dumped back onto the salar.

Approximately 25 samples were collected during the program. The analyses resulted in a map showing the distribution of lithium over the salar; data showed that the lithium concentrations increased toward the south, and north of the Chepica Peninsula. This map served as the basis of the negotiations with CORFO about the extent of the area that the government was willing to cede to Foote Mineral Company. Foote Mineral Company was negotiating for the rights to the whole surface of the salar. However, since Foote Mineral Company was already exploiting the brine deposit at Silver Peak Nevada, there was considerable doubt on the part of the government that Foote Mineral Company would ever develop the salar brine deposit because it would be a significant competitor to its own lithium carbonate production. As it turned out, there was no doubt in the management of Foote Mineral Company that it should develop the deposit because of the extremely high concentration of lithium compared to its Nevada brine and any other known brine deposit. Nevertheless, the Chilean government agreed to cede only 10% of the salar, the area granted to SCL—Sociedad Chilean de Litio—the Joint venture between Foote Mineral Company and the Corporacion de Fomento (CORFO). The joint venture would be controlled 55% by Foote Mineral Company and 45% by CORFO. Foote would fund the feasibility study, the technical know-how and experience, and CORFO would contribute to the minerals rights. Several years later, when the Chilean government realized that Foote Mineral Company was serious about developing the salar as a major producer of lithium, it sold its 45% participation in 1989.

During the early stages of negotiation, Foote had asked for mineral claims over the whole salar. Since the lithium market was well supplied with the Foote lithium carbonate from its brine deposit in Nevada, the Chilean government did not believe that Foote Mineral was going to develop the Salar de Atacama. Therefore, the government asked Foote to select claims that would contain 200,000 MT of lithium (2.34 billion pounds of lithium carbonate) in the southern portion of the Salar de Atacama. So how does one select the claims that the Chilean government would grant Foote Mineral Company?

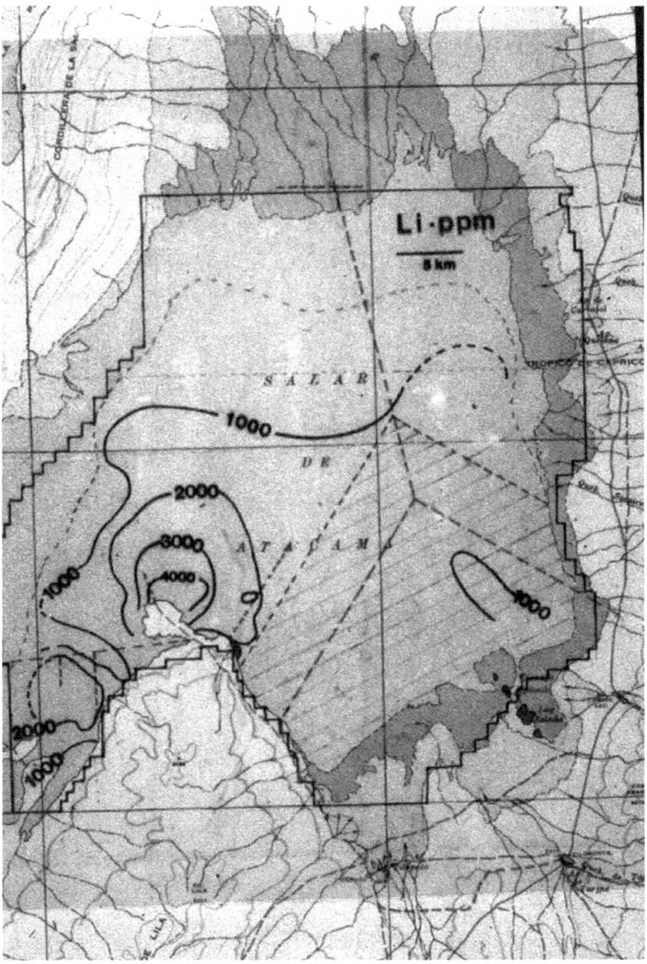

Figure 13.6 Lithium distribution over the Salar based on exploration data.

In 1980, Zenia and I attended the International Geological Congress in Paris with our young children, Markian and Marta. I presented a paper on lithium at the conference. I remember that the chairman of my session wanted me to conclude my presentation quickly because this was July and every Frenchman wanted to get out of Paris as soon as possible to go on vacation. We lodged at the Hilton-Montparnasse. At 3:00 am, Paris time, I get a call from Chile and Phil Walsh, our brilliant negotiator asks me to take out the map of the Salar de Atacama and within one hour call him back with a recommendation on the best claim area in the south of the salar, which would grant a production of 200,000 tonnes of lithium over the life of the project. There I am, half asleep, maps spread over the bed with the children asleep in one bed and my wife next to my maps, and I frantically try to determine the future of the Foote Mineral Project (now controlled by Albemarle). Although I was very familiar with the brine deposit, the highly technical and scientific analysis was not performed under ideal conditions. In retrospect, my knowledge of brines in general and my experience at the Salar de Atacama have indeed resulted in the continuous production of lithium carbonate from 1984 until today, with ever-increasing production rates. The government had granted SCL a total of 200,000 MT of Li production from the SCL claims (2.34 trillion pounds of lithium carbonate). The government thus granted 3,343 hectares, and an additional 2-kilometer buffer zone of 1,370 hectares as a safety corridor against future brine pumping from the northern claims. In retrospect, the SCL joint venture is still in operation in 2020 and going strong and has consistently increased its production.

One of the most interesting "discoveries" during our helicopter exploration was an unusual single trail on the western side of the salar. My further interest was prompted by the remains of longhorn cattle, which are not native to this part of Chile. Usually, we would find donkey bones, but longhorn skulls and horns? This led to a connection with the extensive nitrate industry located further west at Maria Elena which dominated 80% of the world's nitrate industry.

Maria Elena is named after Mary Ellen Comdon, the wife of the first saltpeter refinery works (oficina salitre) manager, Elias Cappelens. Originally, the name of the works was "Coya Norte." Situated on land purchased from the Treasury in 1924, the plant was opened in 1926 using a sodium nitrate extraction system patented by the Guggenheim Brothers which had replaced the Shanks system. In 1965, both were taken over by the state, through the SQM (or "Chemicals and Mining Company").

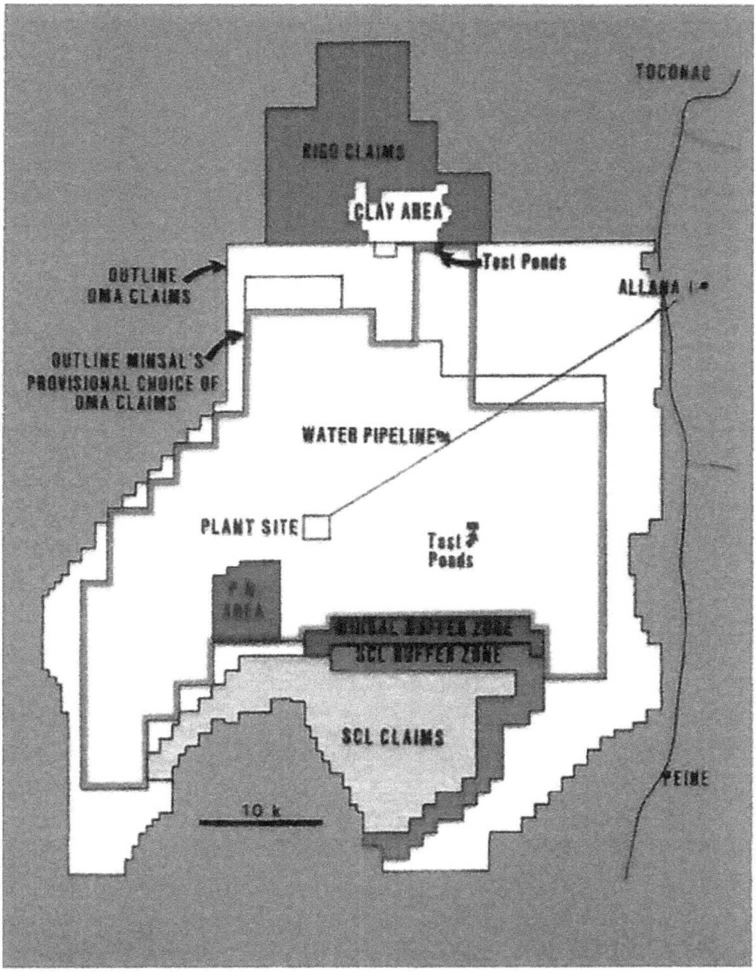

Figure 13.7 Concessions granted to the SCL Joint Venture by the Government of Chile.

Also known as "white gold" or "Chile saltpeter," sodium nitrate is used in the production of fertilizer and explosives. So valuable was this white gold that it caused Chile to go to war with its neighbors in

1879. In the oft-forgotten War of the Pacific or "saltpeter war," Chile, Peru, and Bolivia all fought over territory containing the minerals with Chile gaining much of the disputed land.

Figure 13.8 Single track on the western side of the Salar—presumably used by cattle delivered to the nitrate operations.

The nitrate deposits are in the northern part of Chile mostly in the provinces of Tarapaca and Antofagasta. Since the Atacama Desert, being one of the driest deserts in the world, had little agriculture or cattle industry, food for the workers of the nitrate workers had to be imported. The establishment of the Buenos AiresMendoza railroad in 1885 ended the lengthy and costly trade with carts that connected these two regions of Argentina and facilitated cattle exports from the pampas to Chile, albeit in the last portion of the route the cattle had to walk over the high mountain passes of the Andes. The trail identified during exploration suggests that cattle were driven vis the Socompa Pass east of the salar, down through the village of Socaire then to Tilomonte, and finally to Tilopozo where the cattle could drink their fill at the natural spring before continuing their grueling trek to the nitrate operations. Looking at the trail, one could only imagine the poor first cattle that had to walk over the treacherous razor-sharp salt pinnacles.

Once we started working on the project several activities had to fall into place: the expected flow of the brine to satisfy the original production goal of 12,000,000 pounds of lithium carbonate. This was about the standard production level which would not disrupt the lithium market. One must remember that we are talking about the 1970s when the lithium industry was coming out of the post-

AEC hydrogen bomb program and the lithium usage in pyro-ceramic and grease applications was expanding. At that time, the lithium market was controlled by Foote Mineral Company and the Lithium Corporation of America, sharing the approximate 55-million-pound global market.

Figure 13.9 Core illustrating the high permeability of the salt.

We designed a well-production drilling program that consisted of only two 6-inch production wells. Our data suggested that the upper layer of the salt deposit was very porous. We drilled a test well to a depth of 30 meters (100 feet) and, upon pumping at the very high pumping rate of 1000 gallons per minute, the brine level declined only by 70 centimeters (about 2 feet)!! It seemed that we were pumping from an open brine lake. Indeed, the whole production of 12 million pounds of lithium carbonate required only one well from the high sulfate zone and (discussed below) two high calcium wells.

Since the chemistry of the complex brine needed to be tackled because of its higher magnesium content, a solution had to be designed to prevent the lithium to be lost to complex salts upon the concentration of the brine by natural solar evaporation. The use of lime to process the Silver Peak brine prompted to evaluate the limestone deposits around the village of Peine, deposited as hot springs and to be roasted in kilns to produce lime for mixing with the salar brine and eliminate most of the magnesium and sulfate as Foote Mineral Company did at Silver Peak. While more expensive because of the higher 6.5 ratio of magnesium-to-lithium, an economic evaluation of this option was needed. It turned out that the magnesium would not be an issue in the process as compared with the Silver Peak brines.

Figure 13.10 Extremely high brine flow (1000 gpm).

During the general exploration program, it was discovered that in the promontory separating the salar into two sectors, the Chepica Peninsula was in fact composed of dolomite—a mixed calcium–magnesium rock. The result of this formation was that, after millennia, perhaps millions of years, this formation had "leaked" calcium into the southern peripheral alluvial deposits ringing the Chepica Peninsula creating a pool of calcium-rich brine. The darker alluvial fan surface represented the high calcium sector. The process solution was in front of our exploration group. We drilled 15 exploration wells on the Chepica alluvial fan and found a large resource of calcium-rich brine, which, when mixed with the sulfate-rich brine of the salar provide the solution for the process. Mixing calcium-rich Chepica brine with sulfate-rich salar brine resulted in the precipitation of gypsum eliminating sulfate as a deleterious element which would have resulted in the loss of lithium by precipitating $KLiSO_4$. This pure Na-K-Mg-Cl brine, could now be evaporated and the lithium concentrated as a chloride to 6% Li by solar evaporation.

Figure 13.11 Original SCL Pond system based on a production of 12 million pounds of lithium carbonate.

Following the identification of rich lithium concentrations in 1975, two production wells, Cl-1 (31 meters) and Cl-2 (60 meters) were drilled along the existing CORFO road to confirm an acceptable supply of brine. The lithium concentration from both wells was 1350 ppm Li and the flows highly favorable. Cl-1 was pumped at a rate of 1000 gpm with a drawdown of only 70 centimeters, indicating that the upper salt crust was extremely porous and permeable. Both wells were in the high sulfate zone, but only one CL-1 was required to satisfy the initial 12 million pounds of lithium carbonate production capacity.

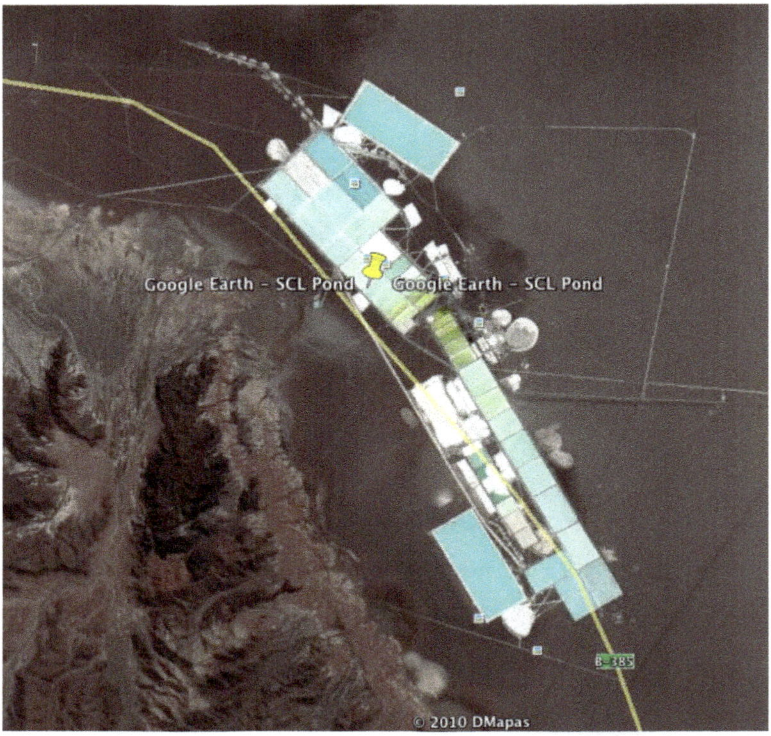

Figure 13.12 Albemarle (new owner of the joint venture) 2010 pond system expansion as a result of increased market demand.

Today, production of lithium has increased dramatically as has the evaporation pond system.

Once the brine source was defined, solar evaporation ponds needed to be constructed. Since the ponds at the Silver Peak Nevada Operation were built on natural clays, the southeastern sector was evaluated for such a purpose. While the permeabilities were satisfactory (10^{-6} cm/sec), the area was overgrown by a plant—the brea—that had a deep root system, which caused the test ponds to leak excessively. Consequently, it was decided to build the ponds on the salar, by crushing the salt ridges and lining the ponds with plastic.

Analytical data showed that the brine was a classical sodium chloride brine with a complex chemistry with lithium, potassium, magnesium, boron, iodine, and sulfate. These elements were input into the closed Atacama structural depression from various sources, volcanic, geothermal, and leached sediments deposited in the depression during the Tertiary, which already was characterized by an anomalous lithium concentration [6].

Handling the complex chemistry required a solution, which would not result in the loss of lithium through precipitation and combination with other elements during the natural concentration by solar evaporation. This was the task of the Foote Mineral team as to how to design a physicochemical system that would accomplish this.

Since the ratio of magnesium-to-lithium in the Clayton valley brine is low at a ratio of 1.6:1, chemists, under the guidance of Dr. Patrick Brown, Director of Research of Foote Mineral Company solved this issue by the addition of hydrated lime which on the one hand precipitated magnesium ad magnesium hydroxide ($MgOH_2$), but at the same time eliminated the sulfate as gypsum. Having eliminated those two impurities, concentration proceeded in the sodium, lithium, and potassium chloride phase. Sodium chloride precipitated first, followed by sylvinite (mixed sodium potassium chloride). The lithium could now be concentrated from 0.025% to a final 4% and be pumped to the chemical plant to precipitate lithium carbonate by reacting with sodium carbonate. However, the magnesium-to-lithium ratio at the salar is considerably higher at 6.7, which would require a much higher amount of lime; nevertheless, I was tasked to evaluate possible sources of limestone that could be burned into lime. A possible source was located under the village of Peine, which sat on an ancient hot spring limestone deposit. My evaluation indicated that sufficient material would be available to react with the brine. However, the economics were not favorable; the process required another solution.

Separating the southern part of the salar into two sectors is the Cordon de Lila ending as the Chepica Peninsula jutting into the salar. Composed of dolomitic carbonate rocks, weathering and erosion deposited substantial alluvial fans which comingled with the brine of the salar. Suspected to contain brines with a high calcium content, 15 exploration wells were drilled along the alluvial fans. 12 of the 15 wells had substantial flows ranging from 43 to 150 gpm; chemical analyses of the brine indicated that, indeed, the brine was rich in calcium, with values ranging from 0.79 to 1.83% Ca, which by mixing stoichiometric amounts of the high calcium brine with the high sulfate brine from well Cl-1 into a pond eliminated the sulfate as gypsum and resulted in a sodium, potassium, lithium, and magnesium chloride brine which could be concentrated to a 6% lithium chloride solution by evaporation. The phase chemistry resulted in the precipitation, as it was at the Silver Peak operation, first of halite (sodium chloride), sylvinite (mixed NaCl-KCl)) and, differently from Silver Peak, the precipitation of carnallite ($KMgCl_6.6H_2O$) and finally bischofite ($MgCl_2.6H_2O$). As the precipitated salts fill up the evaporation ponds, they must be harvested to make room for the next batch of brine and salt precipitation. Magnesium chloride is harvested and stockpiled next to the magnesium ponds and the sylvinite crushed and potassium chloride recovered by flotation.

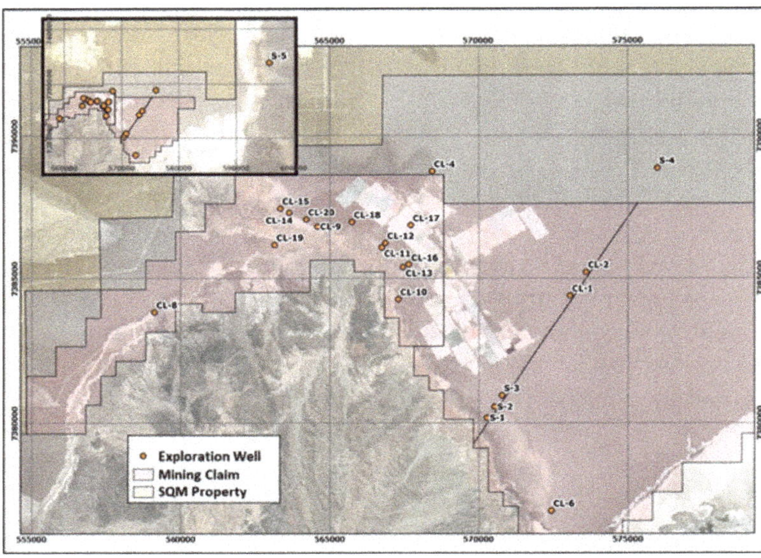

Figure 13.13 Albemarle early well field [9].

The concentrated feed brine was transported by truck some 200 kms from the salar to the la Negra chemical plant site, located east of the city of Antofagasta, where skilled labor and rail transportation were available.

The Andes in northern Chile are separated from the Pacific Ocean by the Atacama Desert and the much lower Coastal Ranges. At those latitudes (18°–26° S) the cold Pacific Humboldt Current and the generally easterly atmospheric circulation accentuate the rain-shadow effect of the western Andean slope to create one of the driest regions on earth. As a result of this, the coastal city of Antofagasta lacks a source of water (drinkable or industrial). It is supplied by pipeline from the lakes high up in the Andes.

Since Antofagasta lacked water, required for the lithium carbonate plant, processing water was delivered by rail from the pipeline delivering water from Andean lakes. As water costs were quite high, I was tasked to locate potential groundwater. A modest drilling program located water with low-enough salinity to replace some of the higher-cost rail water. SCL was thus able to reduce its water costs.

Gene Dezmelyk (with whom I worked on several projects at the salar), vice president of Engineering of Foote Mineral Company designed the lithium carbonate plant. Bob Brennan was the construction manager. The first plant superintendent was Eduardo Morales, who later took over the management of the plant and became the general manager. The general manager of the Sociedad Chilena de Litio was Enrique Arteaga.

13.1 Brine Transport Routes

Besides producing a concentrated lithium chloride solution in the pond system, we had to devise a transportation system to deliver the brine to Antofagasta the port city on the coast. The reason for selecting Antofagasta is that power and qualified labor is not available at the salar. My good friend and professional colleague, Fernando Ide, took off on a three-day trip to explore the feasibility of transporting the brine some 90 kms to the south to the train depot at Neurara, located west of the Socompa Pass; the pass defines the frontier between Argentina and Chile, The rail line is part of the Ferrocarril de Antofagasta a Bolivia (FCAB) and its route through the difficult Andean terrain was designed by American engineer Richard Fontaine Maury, a veritable marvel of engineering in those days.

We packed our camping gear and proceeded to travel south. The terrain was difficult because of many impassable washes, which forced us to drive uphill until we found a passable crossing. I found out that our jeep worked much better in reverse when trying to climb uphill. That night, we set up our tent in a wash and washed our dinner down with copious amounts of great Chilean wine. At night, I had nightmares thinking that a heavy truck would be coming roaring down and crushing our tent. The next morning, we started for the railway spur but found out that the radiator was steaming. We forgot that our trip would take us from a balmy salar elevation of 7500 feet with a balmy temperature (after all it is located on the Tropic of the Capricorn) to an elevation of 1000 feet, where the temperature dropped to freezing. We never thought of using antifreeze. We filled the radiator with our remaining drinking water thinking that we would find water at the Paso Socompa railroad spur. That was a mistake because there was no water available. We turned back and when the jeep finally gave up, we spent the night in the front seats. Stiff with cold, we build a small fire under the radiator hoping to melt the ice, to no avail. We warmed our socks over the fire and proceeded on our 40-kilometer trek with the remaining apple and orange. I had started to walk heroically with my pilot-type attache case but quickly abandoned it. I suppose that some lucky geologist will feel very lucky to discover such an abandoned treasure. We finally got to the crest of the Chepica Peninsula Ridge and saw below the truck that was trying to locate us. Unfortunately, the sun was at such a low angle behind us that we could not use our compasses to signal them. Having no other option, we continued our trek downhill and finally arrived at the flat sector of the salar. Not having had any water for two days and only an apple and an orange between the two of us (it is said that you can only survive 3 days without water in the desert), we were thrilled to come upon a natural spring. Unfortunately, we noticed several bleached donkey carcasses and realized that this was bad water. Walking on was our only choice. Luckily, after a couple of kilometers, we happened to reach a driller's camp, where we could get a drink of water and rest on the driller's heavenly cots. All in all, we had survived the 40 kilometers. Thank God, we were young and had each other for moral and physical support. Our efforts did not result in the selected transportation option. As an alternate route was selected for transporting brine to the Antofagasta chemical plants of Foote at la Negra and SQM at the Salar del Carmen (lithium carbonate and lithium hydroxide). The road starts on the western side of the salar and connects to the Pan American highway in Baquedano, some 40 kilometers from Antofagasta. Although not a straight line to Antofagasta it provided a short distance of 170 kilometers to transport the concentrated brine by tanker truck.

Figure 13.14 Evaluation of alternate brine delivery routes (the author in the company of Fernando Ide of CORFO).

Chile, which, until 1988, was basically a one-commodity country—copper. SCL produced its first pound of lithium carbonate in 1988. When SQM decided to develop the northern part of the salar to produce potassium, the excess brine, rich in lithium, prompted SQM to produce lithium carbonate and become one of the most important producers of lithium chemicals.

After Foote Mineral signed the contract for the feasibility study in 1975, the remainder of the salar, north of the claims granted to Foote Mineral Company remained undeveloped. Several companies, including Great Salt Lake Minerals and Lithium Corporation of America, bid for the exploitation rights. It was in 1986, that MINSAL, created between CORFO (25%), Amax (63.75%), and Molymet, a local Chilean company (11.25%), won the bid. The agreement authorized an expenditure of $300,000,000 over 12 years with the right to proceed or terminate the project after three years from the signing on January 9, 1987. Amax held 67.5% and Molymet 11.25% with the remaining shares.

Under the guidance of Keith Evans, MINSAL undertook an extensive campaign of drilling 12 rotary holes, 34 core holes, and pumping tests in four areas to define the reserves of lithium and potassium more accurately (private report). Results of the exploration indicated potassium (K) reserves of 24 million MT and lithium (Li) reserves of 1.67 million MT. Additional drilling indicated that the basin was indeed very deep [7]. As a result of subsequent drilling to depths of 300 meters, SQM reported total proven and probable reserves of

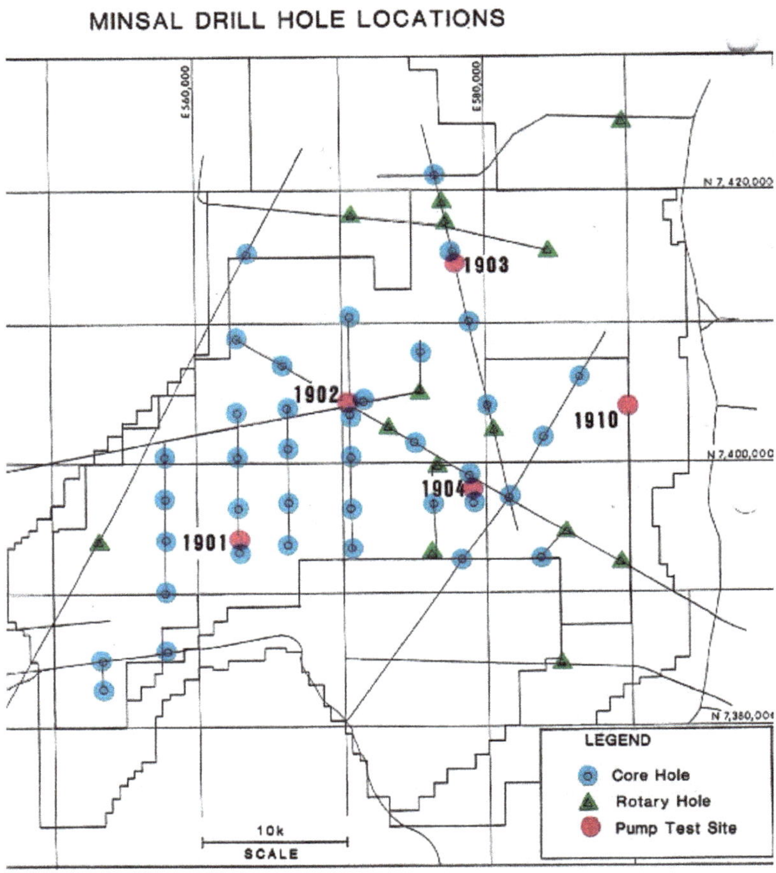

Figure 13.15 Drilling pattern by MINSAL—the first bidder on the Northern Salar claims.

8.55 million metric tons of lithium (as Li) [8]. In 1989, SQM acquired the interest of Amax and Molymet and controlled the bulk of the mineral tenements in the northern part of the salar. In 1995 SQM acquires total control of the northern claims by buying out the interests of CORFO, gaining control of 16,482 hectares and the right to produce 180,100 MT of lithium until 2030. SQM had been formed in 1968 through a joint venture between Compania Salitrera Anglo Lautaro S.A. ("Anglo Lautaro") and CORFO. In 1971, Anglo Lautaro sold all its shares to CORFO, and the company was wholly owned by the Chilean government. In 1983, CORFO began a process of privatization by selling shares to the public and subsequently listing such shares on the Santiago Stock Exchange. By 1988, all shares were publicly owned. On December 21, 2006, two groups of shareholders, the "Pampa Group" (which includes Sociedad de Inversiones Pampa Calichera S.A. and its related companies, Inversiones Global Mining Chile Limitada and Potasios de Chile S.A.) and Kowa Group (which includes the companies Kowa Company Ltd., Inversiones La Esperanza (Chile) Limitada, Kochi S.A. and La Esperanza, a Delaware Corporation) signed a joint agreement and became the controlling group of SQM.

As a result of these transactions, two companies controlled the total rights over the Salar de Atacama to produce lithium: Foote Mineral Company (now Albemarle) in the south and Soquimich in the north.

References

1. Ericksen, G. E., 1973, Geology and Origin of the Chilean Nitrate Deposits, Geological Survey Professional Paper 1188.
2. SQM, 2020, Annual report, www.SQM.com
3. Moraga, A. B., Chung M. E., Fortt, A. Z., 1970, Estudio Geologico del Salar de Atacama Provincia de Antofagasta, *Instituto de Investigaciones Geologicas Boletin* 29.
4. June 26, 1970, Chilean lithium-potassium, *Mining Journal of London*, **274**(7036), 90.
5. Barrett, W., and Kunasz, I. A., 1977, Lithium for thermonuclear reactors, in Considine, D. M. (ed.), *Energy Technology Handbook*, McGraw-Hill Higher Education.
6. Ide, F. Y., and Kunasz, I. A., 1990, Origin of lithium in Salar de Atacama, Northern Chile, in Ericksen, G. E., Pinochet, M. T. C., and Reinemund, J. A. (eds.), *Geology of the Andes and its Relation to Hydrocarbon and Mineral Resources*, Vol. 11, Circum-Pacific Council for Energy and Mineral Resources: Houston, p. 165.
7. Jordan, T. E., *et al.*, 2007, Cenozoic subsurface stratigraphy and structure of the Salar de Atacama Basin, northern Chile, *Journal of South American Earth Sciences*, **23**(2–3), 122–146.
8. Report 20-F and Table 20-2 Lithium Reserves Estimates - Form 20-F United States Securities and Exchange Commission.
9. https: www.sec.gov>data>exhibit 9631231202110-k

Chapter 14

Lithium: Argentina

It is interesting to reflect on the geography of Argentina in evaluating its mining history. Italian settlers who came in droves became important contributors to the economy and culture. In fact, the settlers impacted the way Argentinian speaks Spanish compared to other Latin and South American countries. Activities revolved around the cattle industry and, culturally, this is where the famous tango was born.

Not to be ignored is the great development of the Argentinian wine industry. The birth of renowned Malbec began in the 1800s with a French agronomist Miguel Pouget hired by Domingo Faustino Sarmiento, governor of the province of Mendoza. In the late 1800s, Pouget brought grapevine cuttings from Cahors in France. The province has an ideal climate for the wine industry and its Malbecs are world-famous. Argentina is today the 5th largest exporter of wine in the world.

Although mining of gold dates back centuries, mines were located some 1000 kilometers west of the cultural center of Buenos Aires. Such isolated areas did not attract much interest and it is only recently that important gold mines such as Veladero, have been opened by Barrick Gold in the San Juan province.

There was, however, exploration of salars in those early days. Several scientists [1–3] described the Salars de la Puna, but the focus

The Lithium Legacy
Ihor Kunasz

ISBN 978-981-4968-31-7 (Hardcover), 978-1-003-37236-3 (eBook)
www.jennystanford.com

of their studies was boron; lithium, as it was in the early days of the United States, was not considered.

Today, however, because of renewed interest in lithium batteries for the automotive industry and energy storage, the salars of the Argentinian "Puna" have attracted great interest from several junior companies mostly from Canada in Australia.

BRINES IN EVAPORITIC BASINS, ARGENTINIAN PUNA

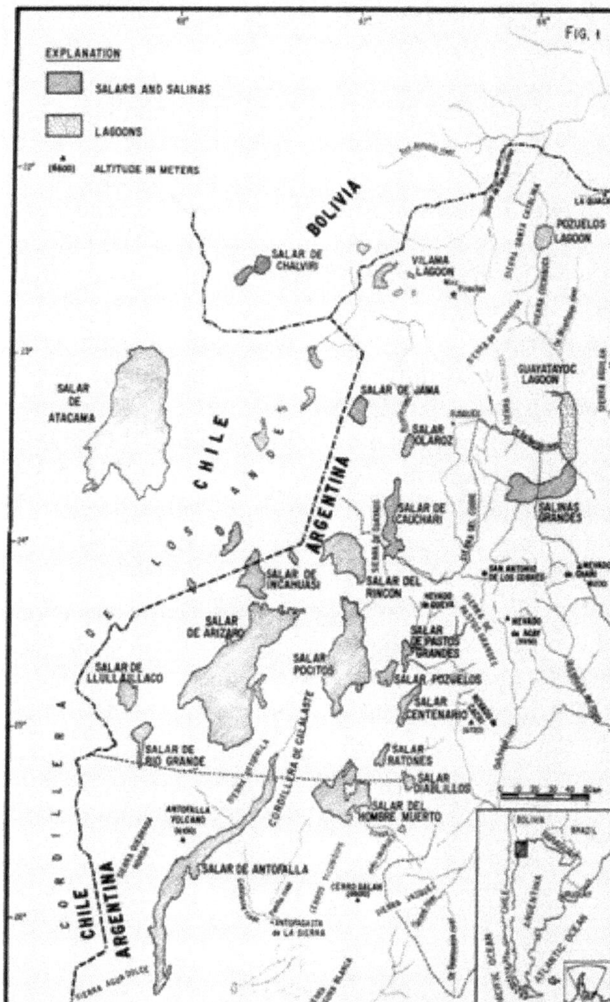

Figure 14.1 Distribution of all known salars in Argentina.

In the 1970s, interest in lithium started, likely because of the lithium development at the Salar de Atacama. Reports by Ericksen [4] and Moraga [5] resulted in an interest in the various salars of the Argentinian Puna.

In July 1972, the topic of lithium was discussed for the first time by geologists of the General Directorate of Military Industries, at which my 1976 publication on lithium on the recovery of lithium from brines was discussed. Lithium became the charge of COPEDESMEL, the Air Force agency responsible for light metals. The evaluation of the lithium potential of the salars was contracted to the National Commission of Spatial Investigations (CNIE). While the program evaluated the chemistry of all salars, special attention was given to the Salar del Hombre Muerto [6].

The evaluation was conducted by Dr. Nicolli, who sampled most of the Argentinean salars. He presented his findings at the International Geological in Paris in 1980. I had presented a paper at the same Congress on lithium brines. Little did I know that I would be invited by Dr. Nicolli to evaluate his data on the Argentinian brines. Samples were collected from 16 salars.

Table 14.1 Composition of brines from various Argentinian salars

Salar	Provincia	$\%SO_4$	%K	%Li	%Mg	Mg/Li
Antofalla	Catamarca-Salta	4.3	0.39	0.03	1.7	56.7
Arizaro	Salta	1.06	0.34	0.07	0.17	2.4
Cauchari	Jujuy	1.5	0.52	0.09	0.2	2.2
Centenario	Salta	4	0.6	0.04	0.3	7.5
Diablillos	Salta	0.85	0.91	0.07	0.38	5.4
Hombre Muerto	Catamarca-Salta	1.2	0.5	0.07	0.2	2.9
Jama	Jujuy	2.2	0.69	0.04	0.07	1.8
Llullaillaco	Jujuy	3	0.85	0.02	0.4	20.0
Olaroz	Jujuy	0.88	0.56	0.09	0.18	2.0
Pastos Grandes	Jujuy	0.9	0.82	0.04	0.39	9.8
Pocitos	Salta	1.5	0.4	0.01	0.22	22.0
Pozuelos	Salta	1.2	0.45	0.03	0.25	8.3
Ratones	Salta	1.1	0.7	0.06	0.34	5.7
Rincon	Salta	4.2	0.71	0.03	0.52	17.3
Rio Grande	Salta	6.3	0.7	0.03	0.42	14.0
Salinas Grandes	Jujuy - Salta	2	0.51	0.04	0.15	3.8

Analyses revealed the occurrence of potentially attractive lithium concentrations in several salars, specifically, the Salar del Hombre Muerto (reminiscent of Death Valley in California) showed great promise and was analyzed in great detail.

In April 1982, Dr. Nicolli sent me a location map of the collected samples as well as the analytical results collected from 26 samples covering the whole salar.

A summary of the analytical data is presented in the following table.

Table 14.2 Summary of Dr. Nicolli's sampling of brines

		Li^+	K^+	Mg^{2+}	Ca^{2+}	$B_4O_7^{2-}$	SO_4^{2-}
Range	From	0.234	2.4	0.023	0.630	0.312	6.4
	To	1.1	11.7	1.7	1.46	1.92	13.7
Most frequent values	From	0.7	7.0	0.5	0.9	1.3	11.5
	To	0.8	8.0	0.9	1.2	1.7	12.5

*Anion and cation concentrations in grams per liter.

Following a cooperative effort between the Argentinean Satellite Institute and the USGS designed to identify salt flats (salares) in Argentina. Dr. Nicolli received the evaluation contract from Fabricaciones Militares, the army agency that controlled critical minerals and metals and undertook the exploration of the numerous Argentinean salars the results of which he published in 1980 [7].

Little did I know at the time of the Congress that I was also presenting a paper on lithium brines and that we would develop a professional relationship. Following several communications, Dr. Nicolli invited me on a trip to evaluate his data and to focus specifically on the Salar del Hombre Muerto, which had shown the highest lithium concentrations.

The invitation was from March 20 to 31, 1982 and involved a very tight program In Buenos Aires. On March 20, a formal meeting took place where a review of the geochemical characteristics of salars was made by CNIE personnel, On Monday 22, I visited the San Miguel Space Center and on March 23, we departed for Catamarca.

I was accompanied by Dr. Jose Suriano, a geologist for the National Center for Space Investigations, Dr. Correa, a chemical engineer for COPEDESMEL, and R. A. Brotkorb, a geologist for the Ministry of Mines. In Catamarca, we were received by the governor of the province after which I gave a paper on lithium resources at the University of Catamarca. We left for Antofagasta de la Sierra and spent the night at the Tincalayu Boron mine of Boroquimica and were hosted by Mr. Raskowski, mine director. The next two days were spent at the Salar del Hombre Muerto collecting samples. I had the opportunity to take 8 brine samples (7 from shallow pits) and one from a well sample from a depth of 29.5 inches using a "guzzler" pump.

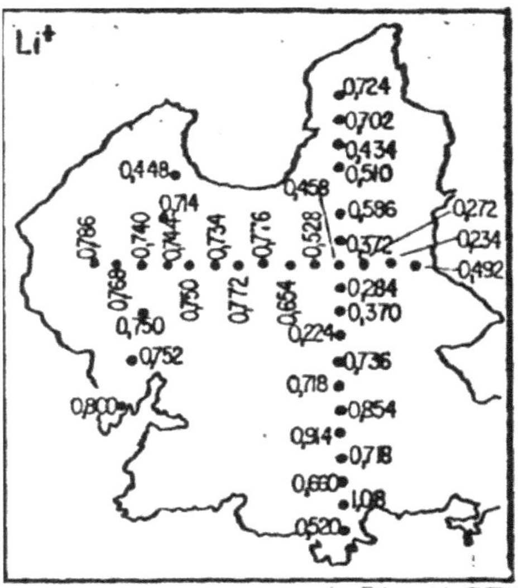

Figure 14.2 Distribution of lithium across the Salar del Hombre Muerto.

Foote's analyses, with minor deviations, confirmed the high lithium concentrations reported by Dr. Nicolli. The lithium values ranged from 450 to 660 ppm, a substantial concentration but well below the values recorded at the Salar de Atacama (1500–2000 ppm Li).

Table 14.3 Salar del Hombre Muerto—Summary of chemical composition

Sample	Li	Na	K	Ca	Mg	Cl	SO_4	B_4O_7	TDS %
HM#1	0.066	10.39	0.623	0.066	0.093	15.26	0.804	0.111	28.57
Hm#2	0.066	10.36	0.61	0.068	0.089	15.67	0.822	0.12	31.16
HM#3	0.066	10.06	0.615	0.059	0.089	15.44	0.611	0.124	30.79
HM#4A	0.063	9.57	0.68	0.065	0.088	15.27	0.828	0.123	27.86
HM#4B	0.062	9.36	0.625	0.06	0.082	15.5	0.814	0.118	27.99
HM#5	0.045	10.12	0.49	0.069	0.082	15.62	0.729	0.058	29.34
HM#6	0.062	9.76	0.577	0.06	0.132	15.57	0.953	0.153	30.5
HM#7	0.051	10.26	0.48	0.068	0.035	15.65	0.759	0.04	29.14
HM#8	0.048	10.07	0.413	0.068	0.06	15.62	0.769	0.094	28.72
Average Foote	0.059	9.99	0.56	0.065	0.083	15.51	0.788	0.105	29.34
Average Nicolli	0.053	10.75	0.51	0.093	0.058	16.17	0.9	0.104	28.14

Foote Mineral undertook an economic evaluation but, since Foote Mineral Company was already involved in a feasibility study to produce an additional 12 million pounds of lithium carbonate at the Salar de Atacama in Chile, Foote management did not see the sense in pursuing two brine projects, especially since the total lithium market in 1980 was approximately 55,000,000 pounds of lithium carbonate equivalents and satisfied by the two U. S. producers [8].

At the time of my visit to the Salar del Hombre Muerto in 1982, I found out that large samples had been collected by US Borax (operating the Searles Lake brine) and an unnamed Japanese company. Lithium Corporation, which would develop the brine had not yet visited the salar.

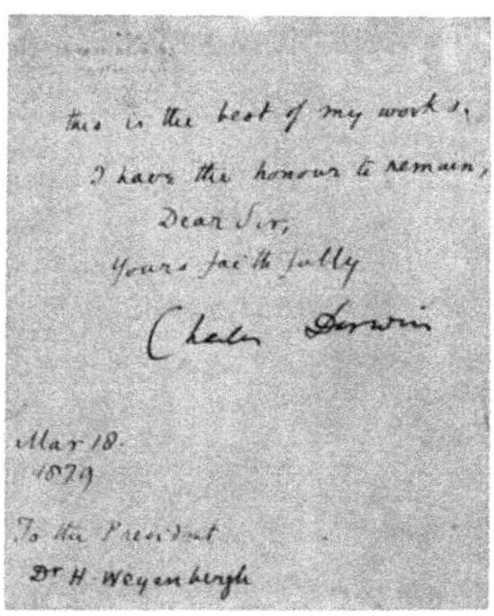
this is the best of my work, I have the honour to remain,
Dear Sir,
Yours faithfully
Charles Darwin
Mar 18. 1879
To the President
Dr H. Weyenbergh

Figure 14.3 Signature of Darwin in the Cordoba Academy of Sciences guest book.

Figure 14.4 Charles Darwin's photo in the Cordoba, Argentina, Academy of Sciences guest book.

Upon our return to Catamarca on March 29, we were received by the Governor of Catamarca, and I was invited to give a lecture (in Spanish, naturalmente!) at the prestigious Academy of Sciences of Cordoba on the "Perspectives of Global Lithium Resources" to an overflowing attendance at the National Academy of Sciences of Cordoba. After my presentation, my biggest surprise was a request to sign the book of prior famous guests who spoke at the Academy. I was told that, unfortunately, the first book was already full and that I would have to sign in the second book. I was shown the first book, and I was pleasantly surprised to see Charles Darwin's signature dated March 18, 1879, when he made a presentation on his famous trip to the Galapagos. I felt honored to be included in such a famous academic company!

In 1997, when SQM decided to produce lithium as a by-product of potassium, and enter the lithium market, it offered lithium carbonate at half the going commercial price of $2.00. This move forced the closure of the high costs of North Carolina spodumene operations. While Foote Mineral Company had its Silver Peak Brine operation, Lithium Corporation of America, was left with no other source of lithium, LCA had been part of the evaluation of the Salar de Uyuni and, anticipating the closing of its North Carolina lithium spodumene operation, needed to locate an alternate source of lithium to satisfy its supply contracts. After canceling its spodumene concentrate contract in 1959 with Quebec Lithium Corporation (which sued the company for $4,500,000) [9], Lithium Corporation then negotiated a lithium carbonate supply agreement with SQM to satisfy their interim contract obligations until it could start production at the Salar del Hombre Muerto in 1995.

In a search for a brine source, Lithium Corporation invested some 5 million dollars in evaluating the production of lithium at the Salar de Uyuni where lithium concentrations as high as 3000 ppm Li had been identified in the southern part of the salar near the Rio Grande delta. That project, however, did not materialize. The Bolivian newspaper *Presencia De La Paz* printed an article announcing that "the (Bolivian) Congress vetoed the project law to grant the concession of the exploitation of lithium of Uyuni." Consequently, LCA left Bolivia. At the time of my visit to the Salar del Hombre Muerto in 1982, LCA had not yet visited the Salar. Yet, U. S. Borax and the Japanese had. To me, personally, the lack of interest was very odd, since Foote Mineral Company was very aggressive in pursuing and evaluating any potential lithium deposit, and found that LCA was somewhat cavalier in their approach.

In November 1990, Flaviano Sabucco, sales manager for South America, placed a call to Buenos Aires expressing its interest in the Salar del Hombre Muerto project. The General Directorate of Military Industries (DGFM) answered positively and suggested several meetings coordinated with the Province of Catamarca. After 3 months of negotiations, an agreement was reached on February

21, 1991. Governor Ramon Saadi requested the approval of the Provincial Legislature, which ratified the agreement on 12 March 1991.

Minera del Altiplano S.A., FMC Argentinian subsidiary, immediately initiated the development of the Salar. It built an elaborate camp and trained Argentineans [10]. It contracted Water Management of Denver to define the hydrology and potential of the salar (Rex Bell, the geologist who later worked with me at the Salar de Atacama did the testing). Updated evaluations now report a reserve of 850,000 MT of Li to a depth of 70 meters. The operation consists of two production centers: the salar proper, where the brine is treated by selective absorption, concentrated in solar evaporation ponds, and then shipped to either Salar de Pocitos where lithium carbonate is produced, and the town of Guemes near Salta where lithium chloride is produced. Rated at that time at 12,000 tonnes per year of lithium carbonate and 5,000 tonnes of lithium chloride, FMC shut down the operation because of technical problems and purchased lithium carbonate from SQM, while the lithium chloride facility continued production [11]. Today the project, now owned by Livent Corp. is the third source of lithium chemicals from brines.

As automotive and storage predictions surged, the exploration for lithium brines demand increased as well. Practically, all the salares listed in Dr. Nicolli report have been explored by a plethora of companies (Alpha Lithium, Advantage Lithium, Alba Minerals, AIS Resources, Dajin, Energi, Eramet, International Lithium, Lithium X, Millenial, Neo Lithium, NRG, Ultra, International Lithium, LSC, POSCO) with hopes of potentially entering the lithium business. However, FMC's Salar de Hombre Muerto project was the only successfully developed lithium brine project in Argentina. This was followed by Orocobre's Salar de Olaroz operation. The Salar de Rincon is still trying to resolve production issues. Minera Exar's (Lithium Americas) Salar de Cauchari is at an advanced stage of development with plans to start production by 2022. Also in the planning stages is Australia's Galaxy Resources Sal de Vida project located east of the Salar del Hombre Muerto. However, a recently announced merger in 2021 between Orocobre and Galaxy would make it the fifth-largest lithium chemical producer [12].

References

1. Catalano, L. R., 1964, Boro - Berilio - Litio (Una Nueva Fuente Natural de Energia), *Estudios de Geologia y Mineria Economica Serie Argentina*, Nro. 3, p. 99. Buenos Aires, Ministerio de Economía de la Nación. Secretaría de Industria y Minería. Subsecretaría de Minería.
2. Schalamuk, I., Fernandez, R., and Etcheverry, R., 1983, Los Yacimientos de Minerales no Metaliferos y Rocas de Aplicacion de la Region NOA (Provincias de Catamarca, Jujuy, La Rioja, Salta y Tucumán), *Anales* XX, p. 206. Buenos Aires, Ministerio de Economía. Secretaría de Estado de Minería.
3. Igarzábal, A. P., and Poppi, R. F. (1980). El Salar del Hombre Muerto. *Acta Geológica Lilloana*, **15**(2), 103–118.
4. Ericksen, G. E. and Vila, T. G., 1978, Lithium Resources of Salars in the Central Andes, Insituto de Investigaciones Geologicas, Santiago, Chile.
5. Moraga, B. A., Chong, D. G., Fortt, Z. M. A., and Henriquez, A. H., 1974, Estudio Geologico del Salar de Atacama, Provincia de Antofagasta, *Instituto de Investigaciones Geologicas Bulletin* 29, p. 56.
6. Méndez, V., 2004, Litio contenido en salmueras de salares de la Puna, Lavandaio, E.O. L., Catalano, E., (eds.), *Historia de la Minería Argentina*. Servicio Geológico Minero Argentino: Buenos Aires.
7. Nicolli, H. B., Suriano, J. M., Kimsa, J. F., and Brotkorb, A., 1980, Caracteristicas Geoquimicas de Aguas y Salmueras de la Puna Argentina, *Miscelana* 63, Academia Nacional de Ciencias: Cordoba, Argentina.
8. Searles, J. P.,1980, *Minerals Yearbook: Metals and Minerals 1980*, Volume 1, Bureau of Mines, U.S. GOVERNMENT PRINTING OFFICE: Washington DC.
9. *New York Times*, Sept. 9, 1959, Lithium Corp. Sued by Quebec Company p. 53.
10. Butts, D., 1992, Personal communication.
11. Ober, J. A., 2000, Lithium: United States Geological Survey, Minerals Yearbook.
12. GlobeNewswire, April 19, 2021, Orocobre and Galaxy agree to a proposed A$4B merger of equals, establishing a new force in the global lithium sector.

Chapter 15

Bolivia: Salar de Uyuni

Although lithium is the dominant topic of interest in this chapter, it was not so in the past. Tin and silver were. Located in the Bolivian tin belt, the Cerro Rico de Potosí is the world's largest silver deposit, with up to 40% silver grades, and has been mined since the sixteenth century, producing up to 60,000 MT by 1996. The famous Cerro Potosí figures on the Coat of Arms of Bolivia.

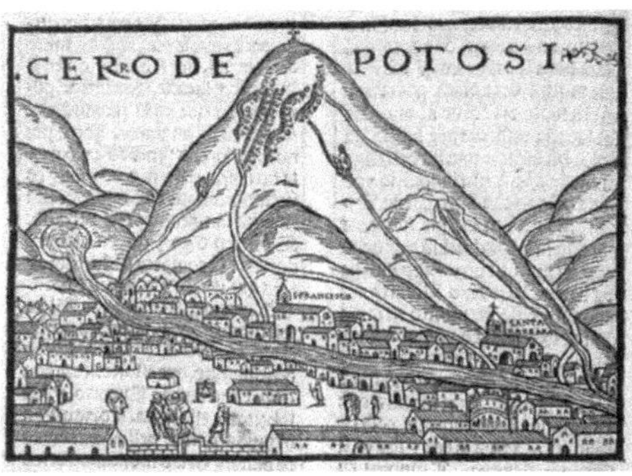

Figure 15.1 Ancient wood-cut print of the Cerro de Potosi, showing early silver mining workings.

Since the colonial period, tin had also been mined in the Potosí region; nonetheless, Bolivia historically lacked the transportation system necessary to ship large quantities of tin to European markets.

The Lithium Legacy
Ihor Kunasz

ISBN 978-981-4968-31-7 (Hardcover), 978-1-003-37236-3 (eBook)
www.jennystanford.com

The extension of the rail link to Oruro in the 1890s, however, made tin mining a highly profitable business. The economy was aided in the late 19th century by a silver boom. When silver prices collapsed, silver production gave way to tin mining. The dominance of mining in Bolivia's economy conditioned the political system. A few wealthy mine and plantation owners, allied with various foreign interests, competed for power. Indians, excluded from the system, found their lot unchanged after almost 400 years.

Figure 15.2 Bolivia's national emblem showing Cerro de Potodsi.

Although the regime encouraged foreign investment, foreign interests and Bolivians with foreign associations took the major share of tin production. This changed, however, when Bolivian tin-mining entrepreneurs realized that smelters in competing countries depended on Bolivian tin. Simón Patiño was the most successful of these tin magnates [1]. Of poor mestizo background, he started as a mining apprentice. By 1924, he owned 50 percent of the national tin production and he controlled the European refining of Bolivian tin. Although Patiño lived permanently abroad by the early 1920s, the two other leading tin-mining entrepreneurs, Carlos Aramayo and Mauricio Hochschild resided primarily in Bolivia.

Figure 15.3 1825 Map of South America after the Spanish Wars of Revolution (note the absence of Bolivia).

Following the 1825 Spanish Revolution, the territory of Bolivia was part of the Viceroyalty of Peru with a long coastal access to the Pacific that included the Chilean Antofagasta and Tarapaca provinces. The 1874 Boundary Treaty, the Treaty of Sucre, abolished the zone of bipartite tax collection on the export dues on minerals found between parallel 23°S and 25°S, Bolivia promised explicitly in article 4 not to increase the existing taxes for twenty-five years on Chilean capital and industry. However, in 1878, Bolivia imposed a new tax on the Chilean "Compañía de Salitres y Ferrocarril de Antofagasta" (CSFA). Chile protested and solicited to submit it to mediation, but Bolivia refused and considered it a subject of Bolivia's courts. Chile insisted and informed the Bolivian government that Chile would no longer consider itself bound by the 1874 Boundary Treaty. On February 14, 1879, when Bolivian authorities attempted to auction the confiscated property of CSFA, Chilean armed forces occupied the port city of Antofagasta in a war that pitted Chile and a Peruvian–Bolivian alliance.

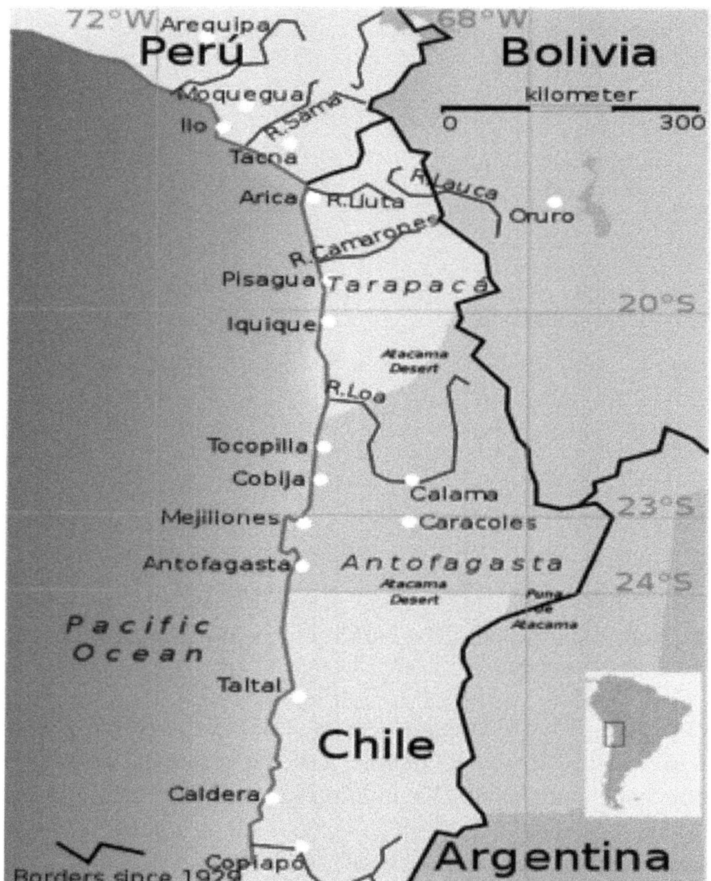

Figure 15.4 Present Map of South America (showing Bolivia).

After the war, which lasted from 1879 to 1884, Chile claimed the total Pacific access, completely isolating Bolivia. Not only was access to the Pacific Ocean cut-off, but also Bolivian access to the previously controlled area rich in mineral deposits (then unknown) such as copper (Chuquicamata and Escondida mines, which produce 50% of the Chilean copper), gold and silver, borax, sulfur, iodine, molybdenum, and now lithium. it can be stated that had Bolivia retained control of those mineral riches, it would have meant a windfall for today's poor Bolivian economy.

The entry of Bolivia into lithium came about through the development of the lithium at the Salar de Atacama, when Foote Mineral company signed a feasibility agreement with the Chilean government in January 1975 to evaluate the possibility of producing lithium chemicals from the brine of the salar.

The first impression in landing in La Paz, Bolivia is the light-headedness because it was the highest airport in the world at the time. Times have changed and I had believed that Lhasa Tibet was the highest, but at 13,323 feet, the El Alto airport is only the second-highest after Chengdu, China with an elevation of 14, 220 feet. At that elevation, the air is quite rarefied and requires special attention during take-off. I often flew with Lufthansa to Santiago Chile with a stopover in La Paz. On one of those occasions, I asked the captain how much of the runway will he use to take off, every **** inch of it!

Bolivia is so different from Chile or any other South American country. Its population is 75% Indian. The women—chollas—wear a mix of colorful clothing including their traditional colorful costumes and, because of the colonial history, skirts still bear witness to the Spanish hoop skirts.

The drabness of the small villages scattered across the Altiplano—the high plateau of the Andes—is compensated by the colorful dresses worn by Bolivian "chollas." In addition, while Bolivian men wear the earflap embroidered alpaca caps, women wear the "Borsalino" hat. The original bowler hat apparently came from England to be used by railroad workers. However, as the workers rejected them, a Bolivian entrepreneur convinced the Chollas that this was the height of fashion in Europe. It worked. The other story I heard in Bolivia is that an Italian hat maker "Borsalino" came to Bolivia to sell his bowler hats to the men who rejected them. It was the women who took to them wearing them with such elegant balance that it is hard to understand how they stay put on their heads. Not only are the hats fashionable, but they also tell a story. Worn straight means that the woman is married. Worn on a slant says: I am available. The majority indigenous population of Bolivia continues with some of the very old traditions such as chewing coca leaves. Chewing coca, "acullico" as they say, is a very common practice in South America. It is done by picking the leaf and putting it between the cheek and jaw along with sodium bicarbonate (or earlier wood ash, which contains potassium bicarbonate). The mixture of the three elements of coca leaf, saliva, and bicarbonate shapes a ball that includes alkaloids and some nutrients. There is evidence that coca was the most widely used plant from pre-historic times in the Andes until today in what are now Colombia, Ecuador, Peru, Bolivia, Chile, Argentina, Paraguay, and Brazil. It is believed that during the colonial period the Spaniards imposed this practice to increase productivity in the mines and reduce their food costs.

Figure 15.5 Bolivian "Cholla"—with her Borzalino hat.

Uninviting though the high, cold country was, it attracted the Spanish because of its rich silver mines, discovered as early as 1545.

Exploiters poured in, bent on quick wealth. Forcing the natives to work in the mines and the *obrajes* (textile mills) under duress, they remained indifferent to all development other than the construction of transportation facilities to remove the unearthed riches. Native laborers were also used on great landholdings. Thus began the system of plunder economy and social inequality that persisted in Bolivia until recent years.

Coca leaf tea is a standard tourist offering, claiming that it will help with altitude sickness, which it does not prevent as my wife, and I experienced in Cuzco. The most lucrative crop is coca, while it has substantially decreased over the years, it still represents about 20% of global production (Encyclopedia.com/places/latin-america).

The Pachamama is another ingrained belief inherited from ancient times. She is the highest divinity of the Andean people since she is concerned with fertility, plenty, the feminine, generosity, and ripening crops, besides providing protection. In the Aymara culture, which lasted from the 5th to the 12th century BC (later invaded by the Incas and the Spaniards) it was very common (and still is today) for the *Pachamama* to receive the first serving of beer at their social gatherings as believers pour a few drops on the ground before they take their first sip, especially at the beginning of a new year.

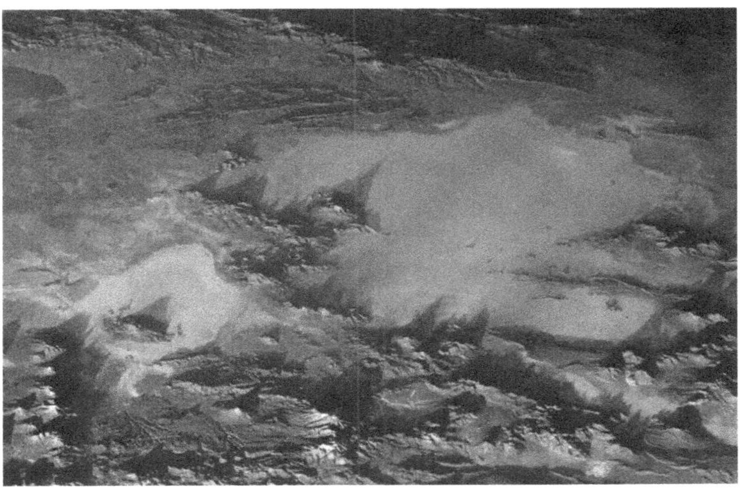

Figure 15.6 Areal view of the expanse of the Salar de Uyuni and neighboring Salar de Coipasa.

At nearly 10,000 square kilometers, the Salar de Uyuni is the largest salt flat in the world. Immensity is the only word that comes only close to describing the Salar de Uyuni. The perpetual silence is only broken by the insistent sound of the wind.

The salar is a relic of a much larger saline lake system—Lake Minchin, named after the person who coined the term—John B. Minchin. The lake covered the areas occupied today by lake Poopó, Salar de Coipasa, and the Salar de Uyuni [6]. The lake had an extension of 200 kilometers (120 mi) in the east-west direction and 400 kilometers (250 mi) in the north-south direction and had depths ranging from 40 to 330 feet [2].

A regional glacial maximum has been associated with the existence of Lake Minchin. Summer insolation was increased during the Lake Minchin period. During that period, precipitation on the Altiplano was higher than today [21]. This precipitation increase started 54,800 years ago, while the time period between Lake Minchin and Lake Tauca featured a dry climate.

Lake Poopó is a neighbor of the much larger Lake Titicaca. During the wet season, Titicaca overflows and discharges into Poopó, which in turn, floods Salar De Coipasa and Salar de Uyuni. While the studies of Risacher and his calculation of lithium resources were calculated for the upper 20 meters of salt, identified at the time of the study, drilling and core analyses confirmed an additional number of salt layers and led to a better understanding of the various climatic phases of the salar [2].

The interest in lithium by Bolivia came about from the successful development of the Salar de Atacama in neighboring Chile. Painful memories of losing the territory of Antofagasta (the Secunda Region), which holds the largest copper, gold, and iodine deposits. This region also includes the richest lithium brine deposit in the world—the Salar de Atacama. At one time, Bolivia even claimed that Chile had drilled under the Andean Mountain chain and tried to drain the lithium brine from the Salar de Uyuni into Chile.

My involvement with lithium in Bolivia came about because my responsibility at Foote Mineral Company included the evaluation of all lithium deposits in the world. The opportunity came about because of the seminal work of Dr. George Eriksen of the United States Geological Survey (George was chief of the USGS mission in Chile) on the salars of South America, which included those in Chile, Argentina, and Bolivia. Although George spent much of his research on nitrate deposits, he contributed enormously to the early understanding of South American salar deposits. His publication with D. G. Chong and G. T. Vila [4] served as the basis for the understanding of brine deposits and their value as sources of important chemical products, such as lithium and potassium.

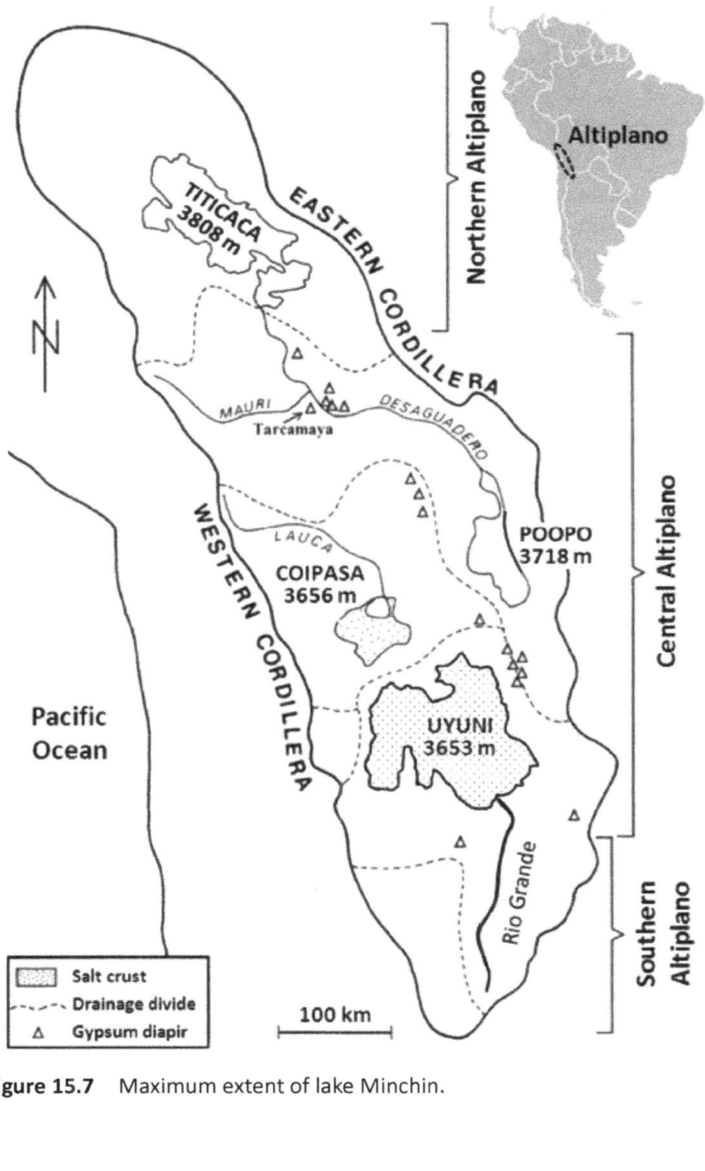

Figure 15.7 Maximum extent of lake Minchin.

The visit to evaluate the lithium of the salar took place in 1976 and looked more like a "lithium safari". Following the identification of significant lithium concentrations in the brines of the Salar de Atacama in Chile, the Bolivian Geological Survey contacted the United States Geological Survey with a request to evaluate the potential of the Salar de Uyuni. The salar had come into focus in 1976 during a Satellite project by the USGS–Bolivian Geological Survey, which identified some highly reflective areas coinciding with highly saline areas in the Salar de Uyuni. Two brine samples collected by W. O. Carter, of the U. S. Geological Survey, Reston, VA, and Raul Ballon of the Servicio de Geologia and Mineria de Bolivia (GEOBOL) were analyzed by Shilrley L. Rettig of the USGS and the report by Eriksen showed concentrations of 490 ppm and 1510 ppm Li [5].

As a result, the Bolivian government invited James D. Vine and George Eriksen from the United states Geological Survey, Gerald Blanton from the Lithium corporation of America and the author from Foote Mineral Company. They participated in a field trip to document the chemistry of the salar brine. Raul Ballon, the representative of the Bolivian Geological Survey, accompanied us. In 1976, the lithium "safari" delegation met at the offices of the Geological Survey of Bolivia (GEOBOL) to discuss the details of the trip. One of the fascinating aspects of the meeting was the handling of the financial contributions by the participants to cover the expenses of the trip. Raul Ballon was shocked to observe that each of us produced the required amount of dollars without having to ask for prior authorization from our superiors. Raul told me later that he would have to get an authorization from the director to purchase even a pencil. After buying the necessary foodstuff, we departed La Paz for the Salar de Uyuni.

Traveling on the Altiplano from La Paz to the Salar de Uyuni, one traverses drab, sad villages where houses are built with gray adobe walls. Here and there, a colorful Cholla pops out of a door breaking the colorless monotony of the landscape. In one of those villages, Caracollo, if I remember correctly, children immediately surrounded us, as is the case over most of the world. Stopping in the only small "tienda," we bought a big bag of candy, which was promptly emptied of its content. I proceeded to buy a second bag when I heard a soft sound from a little boy. "La pelota, senor…la pelota" (the ball, sir, the ball). Having played soccer in France and being an all-American (Western

Reserve University, Cleveland, Ohio). I immediately understood, and there it was, on the wall, probably, in the eyes of the children never affordable, the most beautiful soccer ball in the world. I purchased it and proceeded to the dirt patch, which they called the soccer field. My request for a team picture was immediately and automatically organized according to the standard soccer teams, a row of standing boys and, in front of them, a row of kneeling boys, many without shoes. When I asked about who would be responsible for the ball, a fine young 11-year old volunteered, "I am the club president." Then, I tested their amazing soccer ability but promptly quit after running after the ball at a 12,500 feet elevation. After spending a night at the next village of Challapata, we ultimately arrived at the edges of the Salar de Uyuni.

Figure 15.8 "My" soccer team in the village of Challapata.

As the salar floods during the Bolivian winter, from February to March, there is only one natural access in the southeastern part of the salar where there is a salt recovery operation. Once on the salar, travel was greatly facilitated by the fact that flooding dissolves the rugged and very sharp polygonal ridges, which form in several salars such as the Salar de Atacama. The smooth flat surface allows high-speed traveling.

Figure 15.9 The author standing at 12,000 feet among saguaro cacti.

Figure 15.10 Lithium expedition team collecting shallow brine samples.

As the brine is very shallow, small holes were drilled over the salt flat and each of the participants collected the same sample for analysis at their respective laboratories. It was with great relief that all the analyses were the same for each of the sampled sites. An unusual sight of living in the Sonoran Desert of the southwest United States, what I assumed, was the presence of saguaros. I was surprised to see those at a 12,500-foot elevation in the middle of a salar. The annual humidity during the Invierno Boliviano must be the explanation.

Figure 15.11 Annual flooding on the Salar de Uyuni.

Salt lakes or salars are very treacherous. For one thing, the high elevation of the Salar de Uyuni means the sun radiation is very intense and that the sun reflecting from the salt surface can be very dangerous. In fact, I did not realize that as one walks, one tends to breathe with an open mouth resulting in sunburn on the upper palate, which made for unpleasant sleeping. The other problem is the very quick change of cloud cover. Without any warning, clouds would settle on the salar surface, causing a complete white-out. During a clear day, you have mountain peaks as reference points to guide you. But in white-out conditions, you lose your sense of direction, and you are totally lost. This is precisely what happened two weeks before our trip when a group of archeology students wandered on the salar and became engulfed in the white-out. They drove blindly until their gasoline ran out and were found dead of cold several days later. For any such trips, it is critically important to have a compass, set a direction of travel, and a good odometer for tracking the distance traveled. If bad weather conditions set in, you turn around 180 degrees and backtrack on the exact same path. Luckily for us, the weather was perfect.

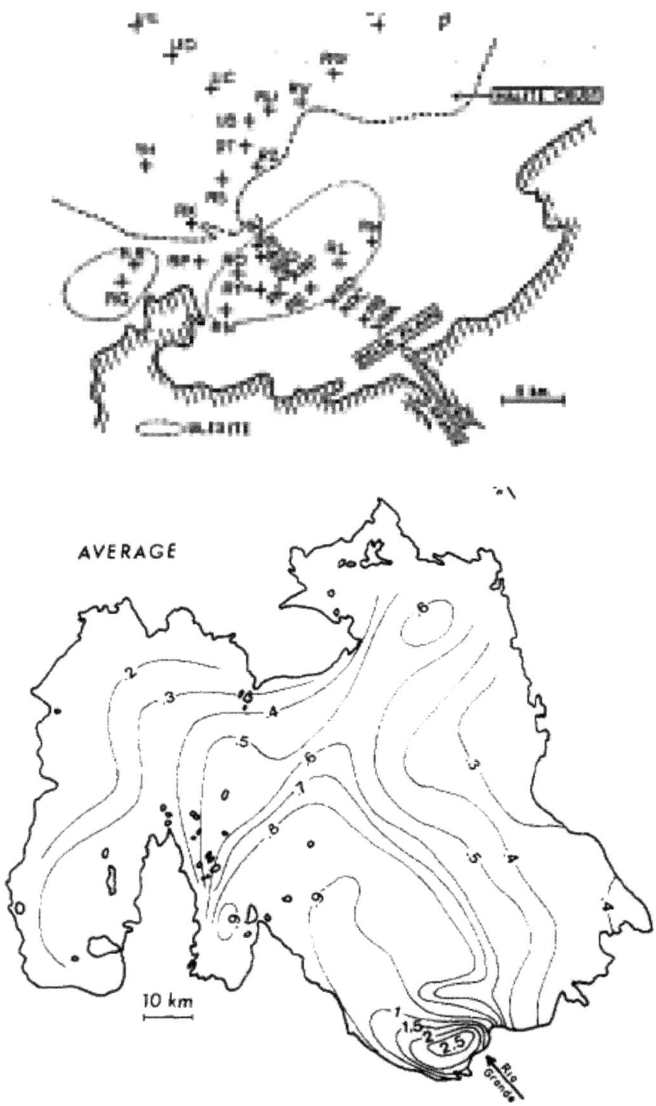

Figure 15.12 Lithium distribution across the Salar de Uyuni based on the sampling by the expedition team (including details of the high concentrations near the Rio Grande delta).

The analytical results of the sampling were published, in 1978 [6]. The resultant map shows the distribution of lithium in samples collected from a depth of 50 cm the depth to at which the brine was intercepted. The values range from a low of 80 ppm to 500 ppm in the main body of the salar. The low values near the edges of the salar are due to dilutions by seasonal floods. Values reaching 2060 ppm Li were recorded in the southern portion where the Rio Grande River discharges into the salar.

After our sampling odyssey, the return to La Paz by train was memorable. When the train arrived at Uyuni around 2 am, the scene reminded one of a second world war chaos when people jumped on any train to escape the onslaught of the Red Army. There was absolutely no ticket control. The local Bolivian travelers were pulled through windows and jumped on top of the cars just to have a chance to get to La Paz. Once on its way, quiet settled in the train. Out of nowhere, I heard some beautiful Altiplano music, with an ethereal sound coming from a zampoña (pan flute) accompanied by the charango, a mandolin-type instrument made of the carcass of an armadillo. It was a group of musicians who were also traveling to La Paz. I offered to buy the group dinner but as the train kitchen was already closed, we settled on a couple of bottles of pisco, the famous grape distillate of Peru and Chile. To this day, there is still a battle between the two spirits, each country claiming theirs is the best.

I had the opportunity to revisit the salar on another occasion in 2005 when the government invited the TRU consulting group to which I belong to propose a lithium-extraction process from the complex brine of the salar. What a difference since the original visit in 1976! The salar had become a very important tourist site. Starting from Uyuni at the southern end of the salar, tens of Toyota jeeps transport thousands of visitors leaving behind indiscriminate tire tracks (and junk) over the salt surface. One can even spend a night or two at a hotel built of salt blocks!

With the available analyses and salt thickness, preliminary estimates indicated that the Salar de Uyuni holds the largest lithium resources in the world. However, as we shall discuss later under brine chemistry, the mere presence of lithium does not make a brine deposit economic. Some elements, which co-exist in brines, can make the extraction of lithium costly and sometimes impossible.

Until 1975, studies of the Bolivian salars had not been undertaken. In 1985, The French Office de la Recherche Scientifique Outre-Mer (ORSTOM), in cooperation with the Bolivian government, conducted a study of the salar. Thirty-eight shallow wells were drilled and sampled. The drilling was done along two sampling lines: East-West and North-South. Results were published in 1989 report titled "Economic Study of the Salar de Uyuni by F. Risacher. Data showed lithium concentrations ranging from 200 ppm Li near the edges of the salar to 800 ppm in the center. Higher concentrations of lithium averaging 1300 ppm lithium and reaching 3500 ppm (potassium concentrations of 2%) were identified in the southeast portion where the Rio Grande Delta discharges into the salar. Then, under an additional cooperation program with the Universidad Mayor de San Andres (UMSA) de La Paz, ORSTOM proceeded to drill several holes, reaching depths of 70 meters. The program discovered nine additional salt layers separated by lacustrine sediments. Each salt layer contains, just like the top layer, a brine rich in various elements. The original lithium resources of 5.5 million tons lithium, based on the upper 20 meters of salt identified in the early shallow drilling program increased to 9.1 million tons as a result of deeper drill information [7]. In 2019, lithium reserves in Bolivia were estimated at approximately 21 million MT [8]. This figure is the result of an assessment incorporating brine in deeper layers.

Drilling and recovery of a 30.74 m core hole, were used to date the sediments accumulated at the salar de Uyuni. The salt collected at 10 meters, show a radiocarbon age of 22,400 years BP, and from 24 to 30.74 meters the age was calculated at greater than 51,000 years BP [9].

Further drilling to depths of 120 meters identified additional layers of salt and clay and increased the lithium resources. A seismic survey experiment was designed, in part, to determine the Quaternary Lake stratigraphy. The seismic profiles were collected using IRIS PASSCAL instruments. These profiles are limited by very low spatial sampling but clearly show that the shallow Quaternary

Lake deposits now dip to the east and north and are offset by sets of faults, particularly on the eastern side of the salar [10].

According to Fritz, three major depositional environments have been identified at the Salar de Uyuni, based on sedimentary structures obtained from cores retrieved from a depth of 220.6 meters.

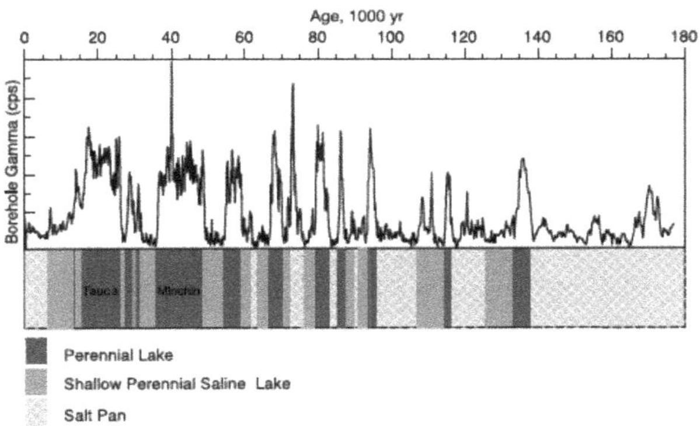

Figure 15.13 Generalized stratigraphy of the upper 100 m of the Salar de Uyuni core versus calendar age.

Strata characterized as "perennial lakes" are mud units from lakes that ranged from shallow to deep and relatively fresh to saline. The mud units vary from massive to laminated and in grain size and concentrations of carbonates (calcite, as mud and as rod-shaped fecal pellets), gypsum crystals, sulfides, organic material, and ostracods and siliceous remains.

Strata characterized as "shallow perennial saline lake" and "salt pan" are salt units that are distinguished based on salt textures. The bedded salts are dominantly halite, with laminae and thin beds of gypsum crystals and carbonate mud and pellets. Halite layers commonly contain well-preserved "chevron" textures outlined by fluid inclusion-banded crystals, as well as sorted crystal plates, rafts, and hoppers, all diagnostic of precipitation in a shallow perennial saline lake [10]. From 100,000 to 180,000 years, the climate was much drier and the deeper halite units are non-bedded massive layers.

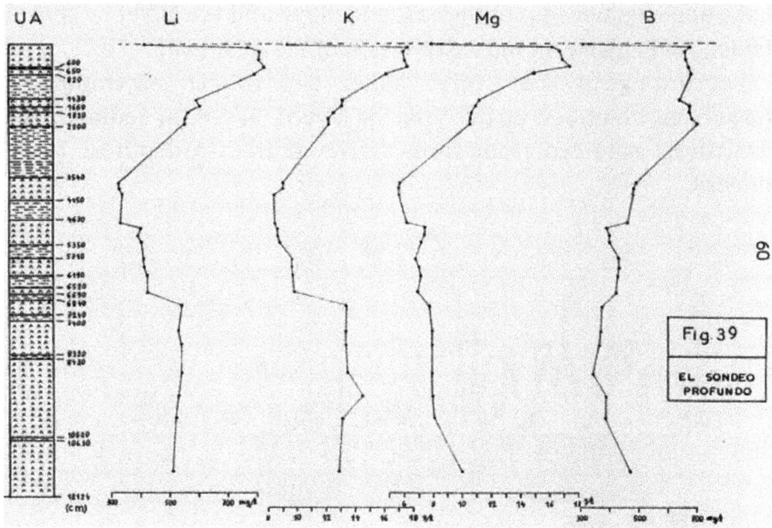

Figure 15.14 Variation of concentration with depth in the 100 m Uyuni core.

As is commonly the case, while salar brines are principally composed of sodium chloride, with subordinate amounts of potassium, magnesium, lithium, and sulfate, the relative concentrations of these elements require different processes to recover economic chemicals. While evaporation in the desert regions where they occur allows for their concentration by solar energy, simple evaporation causes these elements to form as complex salts, from which it is difficult to separate the lithium. The Uyuni brine has one of the highest magnesium-to-lithium ratios of 21.5–26.0 in the shallow brines (20 meters). The magnesium must be removed if lithium is to be recovered . So far, an economic process has only recently been proposed for this site. The process is expensive and makes the brine uneconomic. Uyuni has another problem—it floods. South America is under the influence of counter-cyclonic weather phenomena called "Bolivian Winters." The rains usually come between December and March and can dump as much as 31 inches of rain, which comes down in violent downpours. High flows during the rainy season can breach the retaining dikes resulting in the loss of the concentrated brine lost as well as the loss of a whole evaporating season. This is precisely what happed at the Great Salt Lake in the United States when the Bear River overflowed and tore through the Great Salt Lakes Chemicals Pond system. Furthermore, upon pumping the brine into the evaporation pond system, the accumulated rainwater will sink into the salar and ultimately dilute the existing brine.

This is a risk for the design and operation of evaporating ponds, especially at the mouth of the Rio Grande do Sul where the highest lithium concentration is found. Because of the great potential of lithium to produce lithium batteries to power electric cars, the Bolivian government has been striving to produce lithium from the Salar de Uyuni. Two events presented difficulties in the development of a lithium production center at the salar. One is the very high magnesium-to-lithium ratio of 26. However, analytical data from samples collected in a 120-meter hole show a substantial decrease of magnesium to 8 mg/l while lithium concentration decreased to 500 mg/l, suggesting that pumping deeper brine might provide a better source of brine than the shallow brine considered for production.

Another negative aspect is the intransigent philosophy of President Evo Morales, the first Indian to become president. Although he has been exiled, the new government has continued to impose the same conditions for any foreign investor. The socialist agenda resulted in a substantial delay in developing the mineral resources of the salar. As with many socialist countries, the government wants to create a wholly integrated industry. In this case, the president will not permit any foreign country to simply extract the lithium from the brines and export the chemicals abroad for further processing into value-added products. He wants Bolivia to not only produce lithium carbonate or lithium chloride but to produce lithium batteries and even develop a Bolivian automotive industry.

With the complexity of the brine chemistry coupled with the unrealistic demands of Edo Morales' socialist agenda, the future development of the brines appears complicated, especially since several rich brine deposits with better chemical compositions are in operation in Chile at the Salar de Atacama, where the lithium concentrations are exceedingly high and in Argentina at the Salar del Hombre Muerto; an additional salar—the Salar de Olaroz—has already started production. These deposits together with the spodumene produced in Australia satisfy the lithium requirements for the near-term projected global electric car growth. Recently, in

November 2019, the government of Morales has been toppled and Evo Morales has fled to Argentina. While small-scale production has begun, the future of large lithium production at the Salar de Uyuni is uncertain [12].

Chile and Argentina have far higher-quality reserves of lithium and more favorable climatic conditions for the type of lithium mining carried out in South America. That means they are much, much more appealing as a source of lithium than Bolivia is, at least with current technology.

To extract lithium from brines, mining companies use solar evaporation. That is easy to do in Argentina and quite easy to do in Chile. In Bolivia, by contrast, the Salar de Uyuni is prone to frequent flooding. That means that evaporation takes a lot longer than in any of its neighbors—which adds greatly to the cost of production.

A bigger problem is the extremely high concentration of magnesium—about 20 times as much as in Argentina, and three times as much as in Chile. Current technology requires miners to get rid of magnesium salts by using lime, but that adds to the cost of extraction. Bolivia's magnesium concentrations are about four times higher than the maximum acceptable for cost-competitive extraction, with no easy fix in sight.

Bolivia's poor infrastructure compared to its neighbors also makes operations even more expensive. The salt crystallizes in the dry season, forming millions of tile-looking hexagons that span an area as large as Connecticut. During the wet season, it is covered by a thin layer of water that forms a giant mirror, reflecting the sky so neatly that the line of the horizon disappears. The visual effect draws thousands of visitors and the Dakar Rally every year, making it Bolivia's top tourist destination.

The salt crystallizes during the dry season, forming millions of tile-looking polygons that span an area as large as Connecticut. During the wet seas, however, it is covered by a thin layer of water that forms a giant mirror, reflecting the sky so neatly that the line of the horizon disappears. The visual effect draws thousands of visitors and the Dakar Rally every year, making it Bolivia's top tourist destination.

The pilot plant has been in the southern part of the salar where the Rio grande discharges into it. The reason is that the exploration sampling has identified a very lithium (as well as boron) sector. Its pilot plant in Uyuni has produced close to 250 tonnes of lithium carbonate in 2018, and YLB said it can bring production to 150,000 tonnes within five years. This would make Bolivia one of the top-producing nations and the source of about 20% of the world's lithium by 2022, according to Bloomberg NEF projections.

Figure 15.15 Satellite image of the early experimental ponds in the southeast of the Salar de Uyuni near the Rio Grande Delta.

China's Xinjiang TBEA Group Co Ltd will hold a 49% stake in a planned joint venture with Bolivia's state lithium company YLB. Bolivia estimates that the development of the projects will cost at least $2.3 billion. The Chinese firm will provide initial investment and YLB will pay its share with future lithium production, YLB's executive manager Juan Carlos Montenegro said by phone. Patrick Highsmith, chief executive of Pure Energy Minerals, says he cannot confirm or deny involvement because of Canadian stock market regulations. But in answer to concerns about the extraction processes used, he says his company is experimenting with a lithium-extraction process developed in Israel and particularly suited to the challenges of the Salar de Uyuni. The process involves passing the brine through nanofilters and mixing it with a solvent to separate the lithium from the other minerals. Once the lithium is extracted, the brine is reinjected back into the ground.

Unlike other commercial salar brines currently being processed, the high levels of magnesium and sulfate in Uyuni brine would create difficulties during processing if conventional techniques were used. Two-stage precipitation, like the system adopted by Foote Mineral Company in Clayton Valley, was adopted in the process using lime to remove Mg as $Mg(OH)_2$ and sulfate as gypsum ($CaSO_4.2H_2O$). Boron (at 0.8 g/l in the raw brine), a valuable metal yet deleterious impurity in lithium products, could also be mostly recovered from the brine by adsorption at a pH lower than pH11.3 in this first stage. The residual Mg and Ca (including that added from lime) which were subsequently precipitated as Ca–Mg oxalate could be roasted to produce dolime ($CaO \cdot MgO$) for re-use in the first stage of precipitation. Evaporation of the treated brine up to 30 folds would produce 20 g/l Li liquors. The final precipitation of lithium at 80–90 °C produced a high purity (99.55%) and well crystalline lithium carbonate.

Figure 15.16 Construction of the lithium production facilities by the Chinese.

The political turmoil has not been good for investment in Bolivia. Under Evo Morales, a keen proponent of "resource nationalism," the government would not agree to simply produce lithium carbonate or lithium hydroxide but also lithium metal production, lithium battery manufacturing ending with the production of electric cars, all within Bolivia. With the recent ousting of Evo Morales, there has been no resolution to produce lithium from the brines at the salar. Additionally, local unions have been a substantial negative factor against foreign investment.

There were plans agreed by a privately-owned German firm called ACI Systems Alemania (ACISA) and Bolivia's state-owned lithium company, YLB, to develop a lithium project. However, these resulted in widespread protests as locals said the agreement to build a mine, an electric vehicle battery factory, and a lithium hydroxide plant did not deliver enough local benefits. Eventually, the project was shelved. YLB had also signed an agreement with a Chinese consortium, Xinjiang TBEA Group Co Ltd, for a new $2.3bn lithium project. The company, which beat six others to secure the deal, including Irish and Russian firms, intends to produce lithium and other materials from the Coipasa and Pastos Grandes' salt flats as well following feasibility studies. The new interim chief of YLB, Juan Carlos Zuleta, has already said, as reported by Reuters, that the firm plans to place strict limits on foreign investment in the extraction of lithium. "It is important for the international community to know that Bolivian law says lithium should be extracted and processed by Bolivians," he said in January.

Unfortunately, no progress has been made in developing these enormous resources. The original plan was to produce 30,000 to 40,000 tons of lithium hydroxide a year from 2022, with investments of €300 million to €400 million. Available information indicates that from 13 metric tons in 2017, Bolivia increased and produced 13 metric tons of lithium carbonate in 2017, increasing to 241 metric tons in 2018 and 413 metric tons in 2019. Production was basically halted in 2020, after the failure to reach an agreement with investors in 2019. Lilac Solutions (backed by German BMW), Energy Ventures (backed by Bill Gates), Energy X and CATL (China), Russia's Uranium 1G, China's Gangfeng Lithium, and TBEA have participated in an online meeting with YBL [11]. Recent data indicates that a German consortium has won the rights to the Salar. If the past is prologue, the Bolivian bureaucracy, restrictive laws, and aggressive unions may well further delete a successful project [12]. In fact, on June 15, 2022, the Bolivian government declined to name which of the short list of six finalists it would select.

References

1. Geddes, C. F., 1981, *Patiño, rey del estaño*, A.G. Grupo: Madrid.

2. Fritz, Sh. C., Baker, P., Lowenstein, T. K., Seltzer, G. O., 2004, Hydrologic variation during the last 170,000 years in the southern hemisphere tropics of South America, *Quaternary Research*, **61**, 95–104.
3. Ericken, G. E., Chong D. G., Vila T., Lithium resources in salars in the Central Andes, in Vine, D. (ed.), *Lithium Resources and Requirements by the Year 2000*, U.S. Geological survey Professional Paper 1005, 66–74.
4. Ericken, G. E., Vine, J. D., Ballon, R. A., 1977, Lithium-rich brines at the Salar de Uyuni and nearby salars in Southwestern Bolivia, USGS Numbered Series Open-File Report 77-615.
5. Ericken, G. E., Vine, J. D., Ballon, R. A., 1978, Chemical composition and distribution of lithium-rich brines in Salar de Uyuni and nearby salars in Southwestern Bolivia, in Penner, S. S. (ed.), *Lithium Needs and Resources*, Pergamon Press, 355–363.
6. Risacher, F., and Fritz, B., 1991, Quaternary geochemical evolution of the salars of Uyuni and Coipasa, Central Altiplano, Bolivia, *Chemical Geology*, **90**(3–4), 211–231.
7. Lithium Mineral Commodities, 2020, https://www.usgs.gov/centers/national-minerals-information-center/lithium-statistics-and-information
8. www.nce.noaa.gov, 2022, NOAA/WDS Paleoclimatology - Salar de Uyuni, Bolivia Dtill Hole 50KYR Natural Gamma Radiation.
9. McGeary, S., Bills, B.G., Jimenez, G. 2003, Shallpow Seismic reflection Imaging of the Salar de Uyuni, Bolivia: Quaternary Neotectonics and Stratigraphy, American Geophysical Union, Fall Meeting.
10. Smoot, J. P., and Lowenstein, T. K., 1991, Depositional environments of non-marine evaporites, in Melvin, J. L. (ed.), *Evaporites, Petroleum and Mineral Resources*, Elsevier: New York, 189–348.
11. *PV Magazine*, May 4, 2021, Bolivia launches call for lithium extraction.
12. Mining Technology, 2020, Bolivia: will the ousting of Morales open lithium to foreign Investment? https://www.mining-technology.com/analysis/bolivia-will-the-ousting-of-morales-open-lithium-to-foreign-investment/

Chapter 16

Lithium: People's Republic of China

16.1 Pegmatites

The People's Republic of China does not have substantial lithium resources to compete against Chile, Argentina, Australia, or the Democratic Republic of the Congo. For this reason, it decided to buy raw materials or invest directly in some of the existing operations. However, an enormous amount of research has recently been undertaken and published because of the intense development of the lithium battery industry in China.

The development of lithium in China dates to the 1950s when the Soviet Union processed lithium for their own production of lithium carbonate and hydroxide in Novosibirsk and lithium metal in Krasnoyarsk.

We found out about the Chinese lithium production facilities during a visit to Urumqi in the Xinjiang province of Northwestern China. Foote Mineral Company was invited by the Chinese government to discuss the potential involvement of our company in upgrading their lithium chemical facility. The meeting was arranged through the intermediary of the Japanese Company Marubeni—a lithium carbonate client of Foote Mineral Company. The meeting

The Lithium Legacy
Ihor Kunasz

ISBN 978-981-4968-31-7 (Hardcover), 978-1-003-37236-3 (eBook)
www.jennystanford.com

was interesting as we met a delegation of Chinese and Uygur representatives. This required translation of the comments of Foote Mineral Company vice president, Phil Comer, from English to Chinese to Uygur and back for several hours. Considering today's activities of the Chinese government vis-a-vis the Uygur population, it is quite interesting to reminisce about my discussion with the official people's representative, who spoke French and thus we could carry our conversation in French, and I was told how the Soviets were "kicked" out of the Xinjian lithium mine. The representative told me that the Chinese had resettled 5 million soldiers in Xinjian and then brought in some 2 million young girls from various villages with the stipulation that they could go back home only after 10 years and were encouraged to marry the young soldiers. We saw the dismal original living conditions since none were offered to these new families. They had to dig caves into the gravel terraces and live in them. The overall effect of this migration was that the Chinese could say that there are more Chinese than Russians who were then "encouraged" to return home.

Figure 16.1 Foote Mineral Company delegation to the People's Republic of China (the author is second from right and Phil Comer, vice president of Foote Mineral Co. is in the middle).

As a result of this, the Soviets had to develop their own lithium mine in the Chita Oblast—the Pervomaysky mine in the Zabaikalsky region of the Chita Oblast (which I visited and will discuss later).

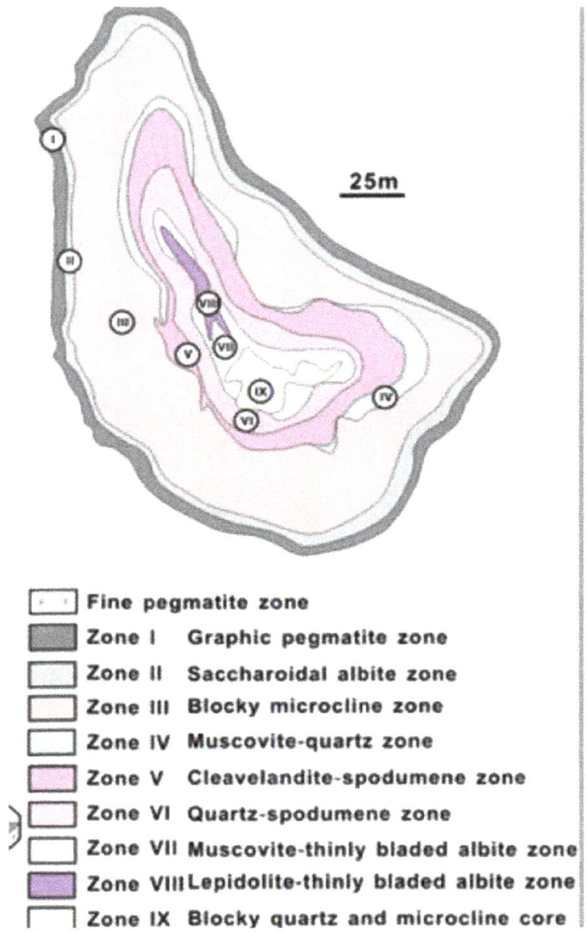

Figure 16.2 Koktokay lepidolite lithium mine.

The visit did not bode well for the Chinese who wanted to find out about the sulfate lithium production technology. When we discovered that the Chinese were still using the old Foote Sunbright, Virginia, spodumene-lime process, Foote's vice president declined any further negotiations which would have included our visit to the Chinese production facility in Urumqi and, in return, a visit by a Chinese delegation to our Kings Mountain, NC facility. Philip Comer realized that we would not gain anything by visiting a plant that used our old, abandoned process and they would gain everything by learning and likely copying our more recent technology. Although we did not get to see the plant nor the mine, the management of the mine gave us several specimens of crushed minerals (presumably from the Koktokay mine and not necessarily representative of ore grades)

- Lepidolite (grade not provided): probably 4–4.5% Li_2O
- Beryl: 10% Beo (BeO)
- Quartz: 99.9% SiO_2

No sample from the spodumene zones (V and VI) was given.

Figure 16.3 Open–pit view of the Koktokay mine.

The Koktokay mine is in the Altai Mountains of northern China in the Fuyun prefecture. The No. 3 pegmatite was the largest Li-Be-Nb-Ta-Cs pegmatitic rare-metal deposit of the Chinese Altai orogenic belt and is famous for its concentric ring zonation pattern (nine internal zones) and rare-metal reserves.

The pegmatite is a classically zoned pegmatite, when concentric zones from the outside shelving progressively evolve to albite, microcline, muscovite-quartz, cleavelandite-spodumene, and finally to an internal lepidolite zone [1].

Mining practices were archaic at best. Some early antimony and tungsten mines were operated by Westerners. After 1940, most of the mines were returned to the Chinese government. However, the Cultural Revolution led to economic depression, most mines were shut down. Later, as I was told, the government encouraged every Chinese citizen to collect any mineral they deemed interesting and bring it to a "triage" center, where trained geologists would identify and select promising ones recommended for further exploration. It is reported that some 168 different mineral products are mined from 20,000 deposits. What is even more amazing is that some 7,000,000 employees are working in the mining industry. At the Koktokay mine, I was told that 4,000 people are working in the open pit. This strongly suggests heavy manual labor and inefficient production methods. However, times are changing, and the Chinese have decided to buy into many foreign deposits (Copper in the Democratic Republic of the Congo, lithium in Zimbabwe, Chile, and Australia.)

China is the world's largest lithium consumer because of its rapid economic growth and the increased demand for electric vehicles. In 2015, China consumed 86,700 MT of lithium carbonate equivalents. China is highly dependent on imports with 70% of the spodumene concentrate imported from Australia alone (representing 50% of the global total) [2]. As a result of the rapid strides in the development of the lithium battery industry, a great number of institutes undertook detailed studies of the lithium occurrences and possibilities in China. The most comprehensive and up-to-date review of the lithium pegmatite and brines has been summarized on the accompanying map [3].

Based on the extensive data, the publication proposes exploration models for locating and evaluating lithium deposits based on "multi-cycle, deep circulation, integration of internal and external" metallogenic mechanism or "MDIE" and the "five levels + basement" exploration model has been successfully expanded to guide the prospecting work both in the Jiajika and Keeryin pegmatite districts in western Sichuan Province. In 2018, six pegmatites were identified in the Sizemuzu district, with three important veins ranging in widths from 10.85 to 15 meters and outcrops from 223 meters to 1553 meters in length. Ore minerals are mainly composed of spodumene (10%–25%), lepidolite (15%–10%), Rb-bearing microcline (0–30%) with minor beryl, and trace columbite-tantalite. Gangue minerals are composed of quartz (25%–40%), microcline (15%–40%), and albite-oligoclase (25%–40%). The main economic ore is spodumene, with a Li_2O content ranging from 0.40% to 3.60% with an average of 1.50%. The claim by Chinese researchers that the

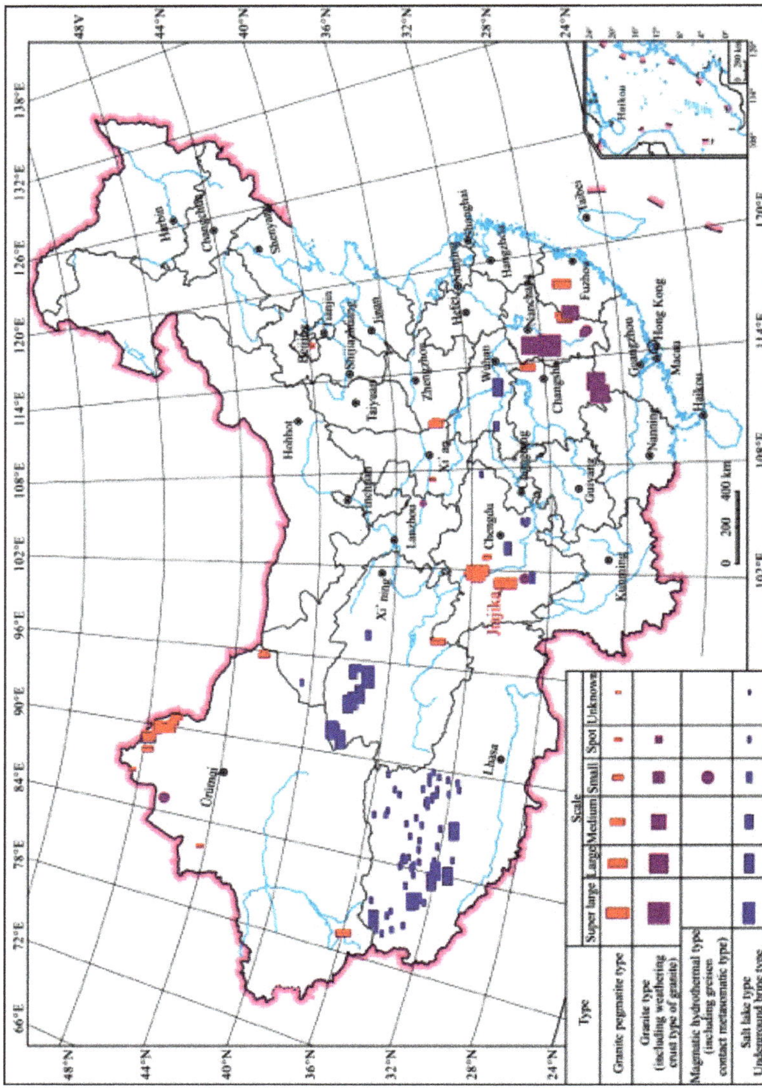

Figure 16.4 Distribution of lithium pegmatites and brines in the People's Republic of China.

deposits, such as the No.134 vein with ore resources (classified a reserves) of 0.512 million tons of Li_2O with an average grade of 1.38% represent a giant industrial grade deposit. The claim that the X03 vein, which is the largest lithium ore body in Asia [3], is in the same class as the Greenbushes pegmatite, Western Australia is unfounded.

From 2016 to 2019, through the implementation of the China National Key Research and Development Program during the "13th Five-Year Plan Period" and by the China Geological Survey project "Comprehensive Investigation and Evaluation of Jiajika Large Lithium Mineral Resources Base in Western Sichuan," seven new Li-bearing pegmatite veins are found in Keeryin, Sichuan Province. Meanwhile, more than 40 pegmatite veins have been newly discovered in Guzhai, Jiangxi Province, expected to host a medium-sized spodumene ore deposit. The significant increase in resources is mainly due to the discovery of the X03 vein, and some new Li-bearing veins have been drilled at the Yakeke area, reporting a total Li_2O resources of more than 2 MT.

Lithium-bearing pegmatites are mainly found in Sichuan, Xinjiang, Jiangxi, Hunan, etc. [4]. The total reported lithium oxide (Li_2O) resource in pegmatites is estimated to be 8.01 million tons, which is equivalent to 18.86 million tons of metallic lithium. These must be understood as raw *in situ* resources, not necessarily reserves available for economic production. One must be careful with the reporting of the lithium or any other mineral or metal resources. The Soviet Union had a great influence on Chinese mineral exploration and resource reporting. Usually "reserves" were reported as total contained elements unsupported by any economic basis. However, my experience with Soviet reporting [I was involved with the development of the Kubaka Gold deposit developed by Omolon Joint Venture (Cyprus Minerals and the Russian Far Geological Expedition)], shows that their resource reporting did not consider economics, cutoff grade, or depth of the vein. If this approach has been used by the Chinese, then the reported resources should be substantially lower.

There is much hope for the lithium industry in China. The small Chinese city of Yichun in Jiangxi hopes to become Asia's next lithium capital by creating and developing lithium related industry worth CNY 100 billion over the next 5 years [5]. The city's mayor believes the metal is in high demand for use in batteries as governments and automakers push to get more electric and hybrid vehicles on the road. To achieve its ambitious goal, the city which had a total gross domestic product of CNY 70 billion in 2009 has created a lithium industry park sprawling over 20 square kilometers and is wooing investors with land use and tax incentives to promote investments to develop the lithium industry to create a complete lithium value chain from mining to automobile battery manufacturing.

But analysts have warned that overcapacity has already started to emerge in China's lithium industry casting a shadow on Yichun's grand plan. They believe Yichun is on a bumpy road to realize its goal. Tibet and Sichuan in China also have large pegmatites reserves and resources in the southeastern province of Jiangxi, but these are largely lepidolite that may prove costlier to extract. It is very hard to grow a complete lithium industry from just a single mine that is not among the best in China.

16.2 Brines

Two main regions of China contain lithium brines: western Tibet (Lake Zabuye) and various brines of the Qinghai Plateau of Northwest China. The earliest summary of the lithium in Chinese brines in the western literature was in 2004 by Don Garrett in which he discusses the brine occurrences in detail [6]. The brine fields tend to be controlled (aligned) by the general tectonic trend of the plateau.

As a result of the dominance of the DRC in the production of lithium batteries and electric car production Much recent and intense research has been conducted in China. Tens of research papers have flooded the scientific publications, addressing the complex chemistries, mainly high magnesium. However, although official information suggests lithium carbonate production levels ranging from 10,000 MT to 30,000 MT.

The following figure is an example of the extensive coverage of brine deposits. A review of the data indicates that the majority represent surface samples likely concentrated by solar evaporation and that the lithium concentrations are highly variable and do not necessarily represent the deeper brines.

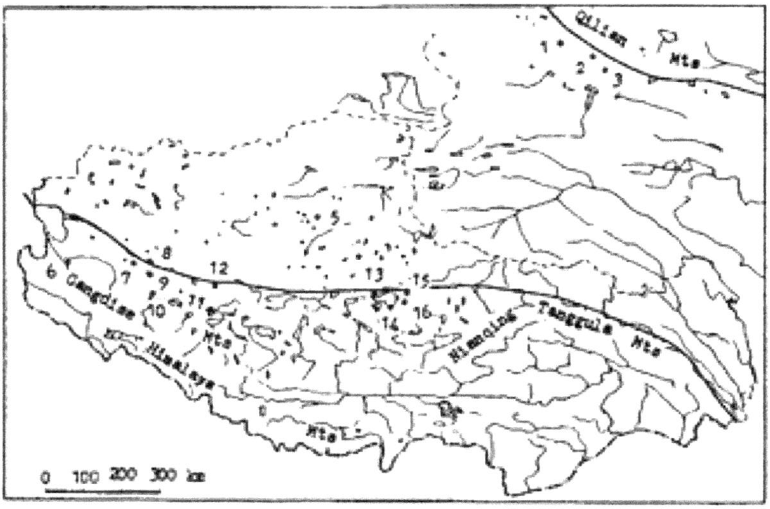

1, Maihai; 2, Da Qaidam; 3, Xiao Qaidam; 5, Pengyan-Co; 6, Geerkunsha; 7, Xia Caka; 8, Zhacang-Caka; 9, Nieer-Co; 10, Calaka; 11, Zabuye-Caka; 12, Laguo-Co; 13, Guojialin; 14, Yagedong-Co; 15-16, Banguo-Co.

Figure 16.5 Distribution of lithium-bearing brines lakes.

16.2.1 Lake Zabuye, Tibet

Although a great deal of literature on the chemistry of the various brines has been published and efforts to produce lithium from the highly magnesian brines of Chaerhan deposit, the only successful lithium carbonate production has taken place at Lake Zabuye, an open lithium-rich saline body. Therefore, a discussion of this particularly unique brine body is of interest and, how does one get to visit Tibet, not as a tourist but lithium geologist to evaluate the reported occurrence of lithium in a lake. The visit was the result of some amazing news that there was an actual lake in Tibet—Lake Zabuye—where high concentrations of lithium had been recorded. The United States National Academies of Sciences showed a strong interest in evaluating this occurrence and invited several well-known academicians to organize a visit and test the lake waters. In the summer of 1988, a delegation of the U. S. Academy of Sciences, consisting of H. D. Holland, G. I. Smith, H. W. Jannasch, A. G. Dickinson, the author, and J. L. Jernow, were privileged to accompany Zheng Mianping and Ding Tiping of the China Association for Science and Technology on an expedition to this lake.

This part of the world has been the result of the collision of the Indian Continent plate and the Asian continent. As the Indian plate slid under the Asian continent, buoyancy lifted the Asian plate, resulting in two continents sitting atop each other and creating the highest mountain chain in the world—the Himalayas. The extreme relief of the Tibetan Highlands was extremely impressive.

Figure 16.6 Members of the National Academies of Sciences traveling to Lake Zabuye, Tibet.

The trip took us first to Lhasa, the capital of Tibet, with a stop at Chengdu. We spent three days acclimatizing to the 12,000-foot-high elevation, a prelude to the next stop at 15,000 feet where we overnighted at a military camp and then finally reached 18,000 feet (where Dr. Holland of Harvard made a show of running a short distance without collapsing) before descending to Lake Zabuye at 14,800 feet.

In Lhasa, we stayed at the relatively new Holiday Inn. New is a relative term. The corridors were lined with buckling carpets, and it was the first time that I saw bedrooms with three beds.

There were two theories on adapting to high elevations: one was to move as little as possible to let the lungs adapt to the altitude or to keep moving so that the lungs adapt more quickly. We opted to walk around the city to get our lungs functioning. One of the memorable visits was to the Jokhang temple—the equivalent of a western cathedral. There, we witnessed how pilgrims would prostrate themselves numerous times in front of the entrance and slowly make their way into the temple. As an aside, it is interesting to note that members of the Eastern Byzantine church tradition, to which I belong, do many similar prostrations during Holy Week. Once inside, the pilgrims delivered their gifts of yak butter, which served to constantly feed the myriads of candles lighting the interior of the many rooms within the temple. Visiting the interior of the temple was fascinating but also difficult because there was no ventilation, the candles exhausted the little oxygen available, and a pervasive strong stench of the yak butter permeated the rooms.

I found Tibetans to be very friendly and trustworthy. In one shop, I did not know how much to pay for an item I purchased. The storekeeper just took my wallet, pulled out the required bills, and returned my wallet. Often the locals would ask us to show them the plastic money used in the United States. We then realized that they were inquiring about credit cards.

We were treated to a visit to the famous Potala Palace—the residence of the Dalai Lama—and to other monasteries. The young monk novices greeted us with smiles and were absolutely fascinated with my chest hair, which they wanted to touch, since, in their genes, body hair apparently does not grow.

Our group was the very first group to ever venture this far west into Tibet (although there were unconfirmed rumors that an Australian had preceded us). The local population was always eager to look at us and, wherever we spent the night, villagers had no compunction to look through the windows of our rooms and just stare at us. In every village, we were awakened by strident music and mottos (I guess, communist) blaring through loudspeakers.

I recall a memorable event in one village. I had taken with me a small tape recorder to play some of my favorite classical music along the way. I noted a curious young fellow watching us. I called him over and put my earphones over his head and played some classical music. I will never forget the bliss in his eyes as he listened to music he had never heard before. All he knew was the blaring of the propaganda music he heard every morning in his village. I was glad to have exposed this young person to some totally different music.

In most villages, we were nicely greeted although the villagers would be naturally shy. As is my habit when I travel to foreign places, I always pack some candy for the children who always surround me. In one village, I recall a parent making sure that the child would thank me. It is then that I realized that in addition to "shie-shie" (thank you), I earned to honor to be addressed as "ye-ye" (grandpa) because of my gray hair.

All along our route, we witnessed the living habits of the native people. They tended to be dirty and dressed in drab clothes. Their tents were low and long and made of yak skins and air. They tended sheep and goats.

They were a curious bunch and always smiling. They were very curious since we looked different, dressed differently, and had equipment unknown to them. At one stop, surrounded by young people, I showed my video camera and invited one of them to investigate the viewfinder and he located a grazing yak. Without telling him, I used my telephoto zoom causing the yak to suddenly come closer. He did not understand the technology and jumped backward at the seemingly attacking yak!

Figure 16.7 Member G. I. Smith surrounded by local children, interested in his camera.

On the return trip, we were giving away our extra rations to the local nomads. They would come back and wanted us to accept their gift of yak butter by the pail full. One of the ethnic delicacies, I would not try again is the Tibetan Tea—a mixture of pungent yak butter mixed and churned with tea.

Lake Zabuye is a landlocked Salt Lake located at an elevation of 4,400 meters (14,400 ft.) in the Shigatse Prefecture of Tibet Autonomous Region, 1,050 km (650 mi) from Lhasa. The lake is surrounded by mountains with a height of 4,600–5,200 m above sea level. It is fed by rain, underground water, and melting ice. The lake gives its name to the mineral zabuyelite (lithium carbonate, Li_2CO_3) Currently, Zhabuye Lithium owns 20-year exclusive mining rights for the Zhabuye Salt Lake. In 1984, lithium was found in micro-fine sediments of the lake and considered amenable to refining in large quantities. The production of lithium from the Salt Lake water started in 2004–2005, after exploration work for the metal was initiated in 1982. Zabuye Lithium has a plant at the lake which had a total capacity of 5,000 MT but produced 1,556.5 MT of lithium carbonate in 2008. Its capacity was projected to increase to 20,000 MT soon. The company claims a reserve of 1.53 million MT Li (8.3 million MT of lithium carbonate), but this estimate is considered overly optimistic.

Lake Zabuye is one of the many high saline lakes on the Qinghai-Xizang (Tibetan) Plateau. It was created during the late Cenozoic tectonic uplift and deformation of the Tibetan Plateau. During the Quaternary period, the lakes covered an area ten times their present extent. Later, moisture from the Indian Ocean was blocked, the region became more arid, the lakes shrank, and many became highly saline. The larger extent of the lake is obvious as witnessed by the many residual desiccation terraces and beaches as the lake dried to its present dimension. While the source of the lithium and other elements is unknown, a spring that discharges into the lake seems likely although it is not clear whether its composition reflects recycling and mixing with saline lake water.

Lake Zabuye, or Zhabuye Chaka, lies at an elevation of 14800 feet a.s.l. It is an alkaline chloride–sulfate brine. It comprises two areas—to the north the lake contains brine and to the south, it is partially a lake and partially a salt flat. Our purpose was to sample the Salt Lake waters and confirm that, indeed, it had an elevated lithium concentration. Not knowing the depth of the lake, we had brought motorized zodiacs, but quickly realized that the lake was too shallow, and the engine just sank into the mud. We had to use our backup inflatable boats and proceeded to collect samples at regular intervals in both the lakes, which took back to the United States for analysis.

Figure 16.8 Sampling of Lake Zabuye brines, the author is on the right.

Zabuye is highly unusual because it is an open brine with an unusually high lithium concentration of 972 ppm Li in the northern lake; the southern drier lake has a concentration of 625 ppm Li [7].

Most lithium salars are dry and salt-encrusted. But, at Zabuye, evaporation of the open brine body results in the natural precipitation of a unique new lithium mineral, **zabuyellite,** with an empirical formula of $Li_2(CO_3)$. This salt has not been identified in any other saline deposit.

The chemical analyses confirmed that the Zabuye brine is an alkaline chloride–sulfate brine with a high lithium concentration of 972 ppm Li in the northern lake and 625 ppm Li in the southern drier lake [7].

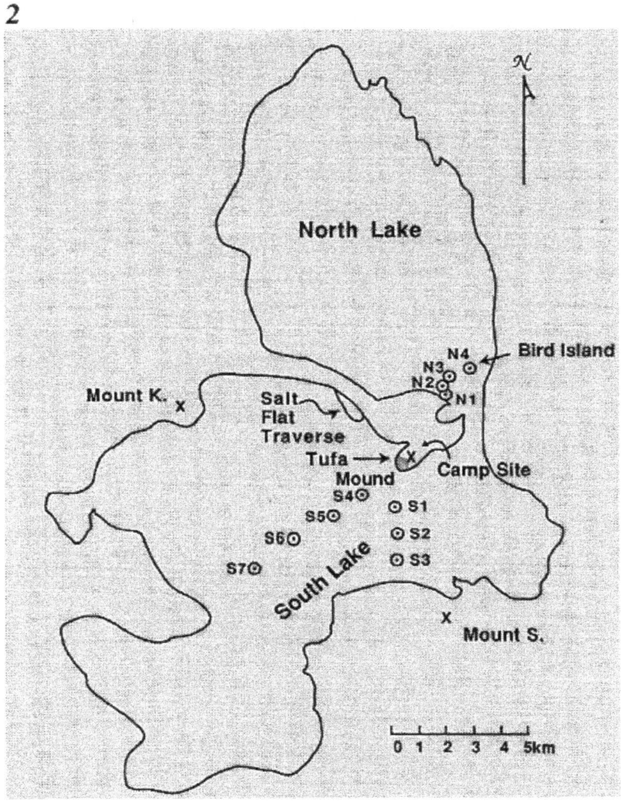

Figure 16.9 Sampling traverses in the southern and northern lake.

16.2.2 Qaidam Brines

Following the visit to Lake Zabuye, Zheng Mianping of the Qinghai Salt Institute invited the author and Joel Christopherson, Manager of Foote's Silver Peak operations to present a lecture on lithium, followed by a visit to Golmud, the site of the major brine operation and a few of the other deposits.

The Shanghai Nonferrous Metals Network, 2019, reports that Zangge Holdings Co. Ltd. (Tibetan Lithium Industry) has an annual production capacity of 20,000 tons of lithium carbonate but will produce about 10,000 tons. This move has been caused by the low lithium and high magnesium contents of the Chaerhan Salt Lake.

The Qinghai plateau in the north of China is the location of some 33 saline lakes, one of which is presently in operation, producing mainly potassium salts. The brine deposits follow a northwest trend. While the various salt lakes of the Qinghai plateau appear to follow a major ESE WNW structural trend, these basins do not conform to the structural ad geochemical signature [3] of other salt basins and appear to be residual from larger bodies as exemplified in the attached cross-section of Lake Taijinaier.

Section of Line 26 of E. Taijinaier Lithium Mine

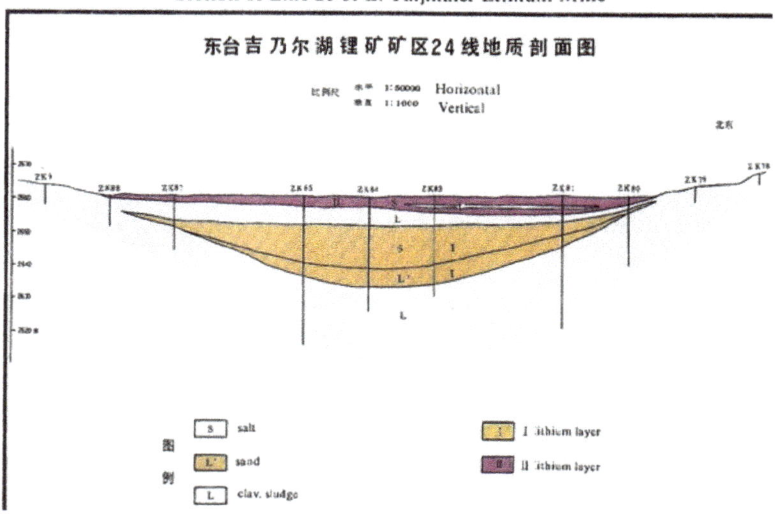

Figure 16.10 Cross-section of Lake Taijinaier, illustrating location of lithium-bearing layers (the author's private information).

In 1993, the author and Joel Christopherson, manager of the Silver Peak Brine operations, were invited by the Chinese Salt Lake Institute to visit the various brine locations. We had the opportunity to visit the main brine operation, located some 60 km from Golmud, which showed a vast pond system of about 10 km^2. Lithium recovery has been hampered by the very complex chemistry, mainly the magnesium content. The operation has only been able to produce about 200,000 try KCl per year. The brine is delivered to the pond system via a system of trench canals from Lake Dabusan, a 96 km^2

dry lake, delivering 27,000,000 m^3 of brine at a concentration of 0.37 g/l LiCl (Chinese often mention lithium but do not specify in what ionic form). Conflicting reports are surfacing indicating, on the one hand, that, because of the complex chemistry at the Chaerhan Salt Lake, the lithium carbonate production will be handled by the Lake Zabuye, Tibet operation (Shanghai Non-Ferrous Metals Network). On the other hand, a 2021 report by the China New Agency, mentions that the Fozhao Lanke Lithium Company will resume the production of a 20,000 ton/year of battery-grade lithium carbonate project in the fall of 2021. The report also indicates a pond surface area of 5,856 square kilometers, (less than the area of 10,000 square kilometers reported during the author's visit in 1993). Also, the chemical data is often questionable as resources and concentrations are reported as lithium Li whereas the actual compound is lithium chloride, a 6.11 higher factor.

16.3 Conclusion

While an enormous amount of research has been funded by the Chinese government, it has become a major player in the production of batteries and electric cars. As these require substantial quantities of lithium, China has realized that it does not have the mineral/ brine base to supply the requirement and that it will not be able to compete with the large pegmatite fields of Australia and the Democratic Republic of the Congo nor with the rich and sizable lithium brine deposits of Chile and Argentina. The government is realizing the weak position of China as a producer of lithium raw materials and has decided to either acquire an interest in existing producing districts or secure long-term supply contracts. No amount of research can create economic deposits.

References

1. Zhou, Q., Qin K., Tang D., Wang C., Tian Y., Sakyi, P. A., 2015, Mineralogy of the Koktokay Numer 3 Pegmatite, Altai, NW China, *European Journal of Mineralogy*, **27**, 433–457.
2. Hao, H., Liu, Z., Zhao, F., Geng, Y., Sarkis, J., 2017, Material flow analysis of lithium in China, *Resources Policy*, **51**, 100–106.

3. Deng-Hong Wang, et al., 2020, Research and exploration progress on lithium deposits in China, *China Geology*, **3**(1), 137–152.
4. Ministry of Natural Resources, China Mineral Resources, Non-oil and gas mineral resources, pp. 7–12.
5. electrive.com, April 22, 2022, The Chinese battery manufacturing CATL to start mining lithium in Yichun.
6. Garrett, D., 2004, *Handbook of Lithium and Natural Calcium Chloride*, Academic Press.
7. Holland, H. D., Smith G. I., Jannesch, H. W., Dickson A. G., Zheng, M., Ding T., 1991, Lake Zebuye and the climatic history of the Tibetan plateau, *Geowissenschaften*, **9**, 37–44.

Chapter 17

Lithium in Medicine

17.1 Bipolar Disorder

Our bodies contain an electrolyte that regulates our electrical and emotional impulses. Therefore, when we sweat a lot, we lose a lot of sodium and potassium, very likely in different amounts. Therefore, sportspeople drink electrolytic liquids such as Gatorade. According to the National Institute of Mental Health, some 9,000,000 people in the United States struggle with the disease at some point in their lives. In 2003, 4.7% of 18–29-year-old adults were afflicted [1].

The behavior of happy/sad is controlled by the ratio of potassium and sodium in one's body. This ratio varies daily and caused us to feel energized up or down. However, when the ratio changes dramatically, we become over-energized for an extensive period and then as suddenly to the point, sometimes, of wanting to commit suicide. The miracle drug, which resolves this problem and brings the ratio into the normal balance is lithium, given to the patient as lithium carbonate. Lithium carbonate is a mood stabilizer. Several companies make lithium carbonate; however, this is "chemical grade" which is the primary chemical to produce many downstream lithium chemicals (over 70). The chemical grade lithium carbonate is sold to the pharmaceutical companies, which upgrade it to a medicinal quality. The lithium carbonate is given to the bipolar patient and

The Lithium Legacy
Ihor Kunasz

ISBN 978-981-4968-31-7 (Hardcover), 978-1-003-37236-3 (eBook)
www.jennystanford.com

the excess passes through urine, allowing for the correct dosage. Information indicates that there are few side effects.

17.2 Historical Background

As early as 200 AD, Greek physicians were prescribing alkaline baths containing lithium carbonate to depressive bipolar patients. Since that time, lithium carbonate has been used as an effective and widely accepted treatment for manic depression. Lithium carbonate acts upon several different neurotransmitters in the brain, including norepinephrine and serotonin. Researchers don't know exactly how this works, but there is an agreement in the treatment community that it does work to stabilize mood and decrease manic symptoms or prevent manic episodes. Dr. Robert Howland reported that there were 16,000 users of the writing of this book.

History tells us that Lithia Springs was the spiritual, healing, and political center for the southern Cherokee tribes. When the Cherokee migrated into Georgia, sometime in the late 14th century, they believed that Yowa their great creator had guided them to where healing waters flowed from a sacred medicine spring that the Creek Indians controlled. To gain access to the healing powers of the medicine spring water, the Cherokee had to fight the Creek Indian Tribe over control of the healing spring. Instead of warfare, the two tribes agreed to play a Lacrosse game, with the winner taking control of the springs. The Cherokee won the Lacrosse game and with the victory came the right to control the medicine spring and its healing Lithia Water. In 1838, the Cherokee and Creek were removed from Georgia (trail of tears).

Even though lithium was first discovered as a chemical element in 1817, the medical profession was interested in "urate imbalances," which were thought to explain a variety of diseases, including mania and depression. Around this time, it was discovered that a solution of lithium carbonate could dissolve stones made of urate, but not treat any illness. The first recorded use of lithium for the treatment of mania, based in part on the urate/lithium connection, was in 1871. The use of lithium carbonate (the current pill form of lithium) to prevent depression came in 1886. As the public learned about lithium, great interest in this mineral led to the use of mineral-rich spring waters in spas, baths, and beverages. Because most of these mineral waters contained only traces of lithium, the dangers of lithium and higher concentrations were not recognized. When a tablet form was used as a salt substitute in low-sodium diets, there were many reports of severe lithium side effects and some deaths.

Figure 17.1 Cherokee Indians competing in a Lacrosse game.

Just as the dangers of lithium were becoming apparent, an Australian psychiatrist named John Cade, an Australian psychiatrist, began treating patients with mania using lithium (1948). He too was led to this approach from a focus on lithium and urate. He had injected guinea pigs with lithium urate and found that they became placid, and somewhat tranquilized. Only later did he determine that the calming effect was from lithium, not urate.

In 1949, Cade discovered that lithium carbonate could be used as a successful treatment of manic depressive psychosis [2]. Cade's findings did not immediately lead to treatments and many years passed before lithium was in widespread use for the treatment of bipolar disorder. Speculating about why lithium was not immediately adopted by the psychiatric profession, Cade stated that a discovery "made by an unknown psychiatrist with no research training, working in a small chronic hospital with primitive techniques and negligible equipment, was not likely to command attention." However, with careful attention to dosage and blood concentration, the effectiveness of lithium for patients with bipolar disorder was slowly established.

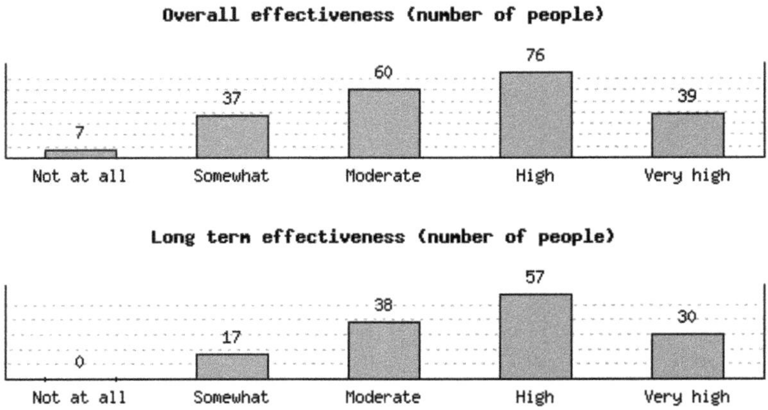

Figure 17.2 Effectiveness of using lithium carbonate in bipolar people.

In the 1950s, U.S. hospitals began experimenting with lithium on their patients. By the mid-60s, reports started appearing in the medical literature regarding lithium's effectiveness. The U.S. Food and Drug Administration did not approve of lithium's use until 1970 it was approved by the FDA for the treatment of mania in 1970.

The impact of Cade's discovery can also be considered at many other levels: the relief of suffering for multitudinous bipolar patients and their families, and the economic benefits to the wider community. It has been estimated that from 1970 to 1994 lithium saved the USA alone over $145 billion in hospitalization costs [3].

17.3 Doctors Around the Country Prescribed Lithia Water for Health

In October 1887, the Congress of Physicians held its annual convention at the site of Lithia Springs, Georgia. Historical records [2] show that prestigious doctors of the day prescribed lithia water for gout, rheumatism, urinary tract disease, depression, mood disorders, and many other ailments. In 1891, Bowden Lithia Springs published a book entitled *The American Carlsbad and Its Famous Medicinal Waters*, which contained the history of Lithia Springs and a full body of testimonials from famous physicians of that era.

Figure 17.3 Congress of physicians, Lithia Springs, New York, 1887.

Above is a vintage photo of the Congress of Physicians 1887 held at Lithia Springs, Georgia. These gentlemen were accomplished doctors and respected as healers at this time in history. They may not have had our modern-day science and technologies, but they had intuitive knowledge and empiricism in knowing what healed. The reason their famous Congress was held at Lithia Springs was that the water that flowed from Lithia Springs had legendary healing powers.

Lithium	0.5 mg/l
Magnesium	8 mg/l
Potassium	17 mg/l
Sodium	450 mg/l
pH	7.3–8

17.4 Famous Lithia Springs Sweet Water Health Resort

In 1888, the Sweet Water luxury hotel and health resort with 250 rooms was opened at Lithia Springs. This first-class health resort was considered 50 years ahead of its time in offering its guests electric lights and intercommunications system (the first in Georgia), indoor plumbing, custom-made furniture from Michigan, European wines, and linens. The total treatment included lithia vapor steam and electric massage on a marble slab to soothe irritated skin, body, and mind, which provided comfort for hours after a treatment. This first-class prestigious health resort hotel was touted as having the most modern-day amenities of electricity and telephones while serving LITHIA™ to each room. A train line ran directly from New York City to Lithia Springs, Georgia, with a stopover in Atlanta. Mark Twain, the Vanderbilts, President Cleveland, Taft, McKinley, and Theodore Roosevelt were just some of the rich and famous people who came to Lithia Springs to drink LITHIA, the legendary health elixir. A miniature train called the "the little Anna" ran from the Sweet Water Hotel to Lithia Springs.

Figure 17.4 Sweet Water Luxury Hotel, Lithia Springs, New York, 1888.

17.5 Lithia Park, Ashland, Oregon

In 1907, a lithia water spring was discovered at Emigrant Creek several miles to the east. Upon analysis, the water was shown to have the second-highest concentration of (presumably beneficial) lithium in any natural spring (the highest being in the famous springs of

Saratoga, New York). Bert Greer, a journalist, moved to Ashland in 1911 and purchased the *Ashland Tidings* newspaper. He agitated for the idea of establishing a mineral water resort at Ashland and campaigned for a bond issue to fund mineral springs-related improvements to the park.

17.6 British Columbia

Happy Water is a new bottled water product containing 0.1 parts per million of lithium (lithia salts) from two natural springs from BC mountain ranges—Halcyon, in the Kootenays, and another on Mount Woodside, near Agassiz. The products are promoted by Smile Squads of young, yoga-type, uber-enthusiastic boosters traveling across the province. They are ubiquitous at events wherever the health-seeking adventurous people go [4].

17.7 Lithium in Drinking Water and the Incidences of Crimes, Suicides, and Arrests Related to Drug Addictions

Using data for 27 Texas counties from 1978–1987, it is shown that the incidence rates of suicide, homicide, and rape are significantly higher in counties whose drinking water supplies contain little or no lithium than in counties with water lithium levels ranging from 70–170 μg/l; the differences remain statistically significant (p less than 0.01) after corrections for population density. The corresponding associations with the incidence rates of robbery, burglary, and theft were statistically significant with p less than 0.05. These results suggest that lithium has moderating effects on suicidal and violent criminal behavior at levels that may be encountered in municipal water supplies. Comparisons of drinking water lithium levels, in the respective Texas counties, with the incidences of arrests for possession of opium, cocaine, and their derivatives (morphine, heroin, and codeine) from 1981–1986 also produced statistically significant inverse associations, whereas no significant or consistent associations were observed with the reported arrest rates for possession of marijuana, driving under the influence of alcohol, and drunkenness. These results suggest that lithium at low dosage levels has a generally beneficial effect on human behavior, which may be associated with the functions of lithium as a nutritionally essential trace element. Subject to confirmation by controlled experiments with high-risk populations, increasing the human lithium intake by supplementation, or the lithiation of drinking water is suggested as a possible means of crime, suicide, and drug-dependency reduction at the individual and community level [5].

There is increasing evidence from ecological studies that lithium levels in drinking water are inversely associated with suicide mortality. Previous studies of this association were criticized for using inadequate statistical methods and neglecting socioeconomic confounders. This study evaluated the association between lithium levels in the public water supply and county-based suicide rates in Texas. A state-wide sample of 3123 lithium measurements in the public water supply was examined relative to suicide rates in 226 Texas counties. Linear and Poisson regression models were adjusted for socioeconomic factors in estimating the association. Lithium levels in the public water supply were negatively associated with suicide rates in most statistical analyses. The findings provide confirmatory evidence that higher lithium levels in the public drinking water are associated with lower suicide rates. This association needs clarification through examination of possible neurobiological effects of low natural lithium doses [6].

17.8 7-Up Drinks and Bottled Water

Lithium drinks were in huge demand for their reputed health-giving properties, so much so that the element was added to commercial drinks. 7-Up was originally called "Bib-Label Lithiated Lemon-Lime Soda" and contained lithium citrate right up until 1950. It has been suggested that the 7-Up came from the lithium 7 isotope which remained after the separation of the reactive lithium 6 isotope, used in the production of the hydrogen bomb.

Since lithium was withdrawn from our water and beverage supply in 1949, America has witnessed not just an increase in autism, mood disorders, divorce, dental decay, dementia, diabetes, drug use, and diseases like Huntington's and Parkinson's but it has seen a host of new problems such as Alzheimer's, ADD, ADHD, OCD, and bipolar disorder. All these problems, social and physical, have grown increasingly with each new generation as we have gotten less and less lithium and a host of other minerals.

Figure 17.5 First use of lithium citrate—Lithiated Lemon Soda by 7-Up.

Scientist Victor Schauberger from Austria, a hundred years ago, told his peers and colleagues that one day in the not-so-distant future, water would be more expensive than wine or gasoline. He was written off as crazy. Many foreign companies listened and have invested heavily in American municipal water supplies and the U. S. bottled water industry. Now Americans think nothing of buying a 12-ounce bottle of water at public events for \$1.00 to \$2.50 which is equivalent to about \$11–27 per gallon of gasoline (depending on the market).

References

1. Jonas, B. S., Brody, D., Roper, M., and Narrow W. E., 2003, Prevalence of mood disorders in a national sample of young American adults, *Social Psychiatry and Psychiatric Epidemiology*, **38**, 618–624.

2. https://en.wikipedia.org/wiki/History_of_bipolar_disorder#cite_note-12
3. Fieve, R. R., 1975, *Moodswing: The Third Revolution in Psychiatry*, Morrow: New York.
4. Fayerman P., 2012, Will drinking lithium in Happy Water sourced from B.C. mountains put a spring in your step and smiles on you face?, www.vancouversun.com
5. Schrauzer, G. N. and Shrestha, K. P., 2010, Lithium in drinking water, *British Journal of Psychiatry*, **196**(2), 159–160.
6. Blüml, V., Regier, M. D., Hlavin, G., Rockett, I. R. H., *et al.*, 2013, Lithium in the public water supply and suicide mortality in Texas, *Journal of Psychiatric Research*, **47**(3), 407–411.

Chapter 18

Lithium Demand and the Electric Car

For many years, the lithium demand had been principally in the ceramic, grease, synthetic rubber, and lithium batteries for the exploding electronic industry. Now, because of the potential for automotive and energy storage developments, lithium projections have mushroomed. The fundamental reason for electrification is the move toward a carbon-neutral and ultimately carbon-free environment. In turn, this has been based on the negative impact of carbon dioxide (today methane has been added as a cause) on temperature and the resultant decision by the United Nations and accepted by most developed countries that world temperatures should not exceed 2°C. The basis of this number is still debated. However, data from past centuries offer actual insight into this issue. Data from the Vostok ice core reveal that the 2-degree maximum has been exceeded as far as 400,000 years ago when there was no industry to cause such an increase. The graph also shows that temperatures have been mostly cooler historically (as low as –8°C). Another critical piece of information is that the carbon dioxide level does not precede (i.e., cause temperature increase), but follows the temperature trend. Finally, a point that is never addressed, is the historical concentration of dust that blocks the sun's heat and is perfectly inversely correlated with temperature and causes temperature and carbon dioxide minima.

The Lithium Legacy
Ihor Kunasz

ISBN 978-981-4968-31-7 (Hardcover), 978-1-003-37236-3 (eBook)
www.jennystanford.com

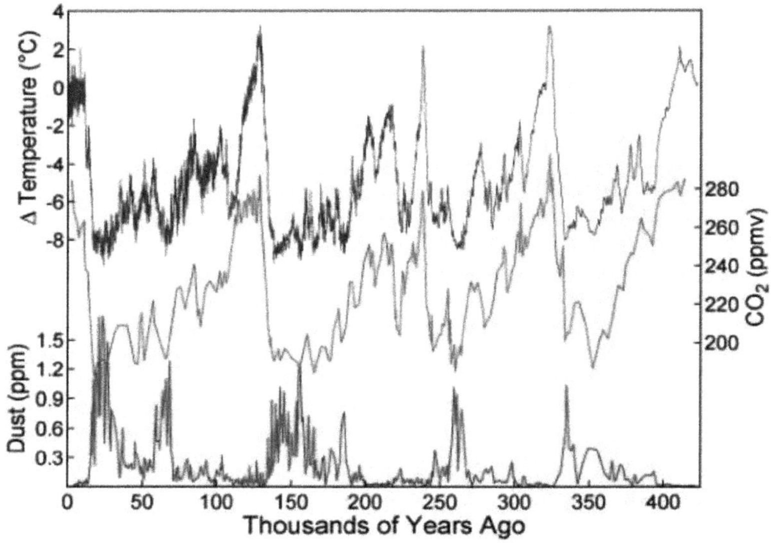

Figure 18.1 Temperature, carbon dioxide, and dust concentrations over the past 400,000 years (based on the Vostok, Russia Ice core).

Using the UN mandate, automotive companies are aggressively moving toward electric cars and optimistic projections are being advanced regarding the future of electrification.

What projections have not evaluated is the availability of not only sufficient lithium resources to sustain an ever-increasing demand, but also the capacity of the global lithium industries to supply such projected demand. This has prompted the United States Government to call for the identification of critical elements, which include lithium. One such evaluation was undertaken by the National Renewable Energy Laboratory, operated by Alliance for Sustainable Energy, LLC, for the U. S. Department of Energy (DOE). The report lists several resources, although the Rhine Geothermal Project and the Argentinian Kachi salar are not U. S. projects [The author should have included the still existing potential of the North Carolina pegmatites of the two presently closed pegmatites and a new project (Piedmont resources) that contain substantial reserves of lithium that could be quickly put back in operation]. Also absent are the substantial lithium resources contained in the brines of the Great Salt Lake, where two companies (Compass Minerals and Great Salt

Lake Minerals) are considering the production of lithium carbonate and the potential from the Imperial Valley Geothermal system [1].

The most recent information on global lithium producers and future annual production compiled by the author is summarized in Chapter 10, based on the most recent company reports.

The total production capacity of the existing producers is listed as 468,000 MT of lithium carbonate equivalents. If the anticipated production goals of existing companies are reached, then there might be as much as 1,117,525 MT of lithium carbonate equivalent available. As there are many producers intent on entering the lithium market, some additional 422,400 MT of lithium carbonate might be available in the future (see Chapter 10 for details).

Roskill predicts that from 2017, when the demand stood at 144,000 MT, lithium carbonate equivalent (LCE) production capacity was 270,000 MT and estimates that the demand will grow to 417,000 MT assuming a compounded annual growth rate of 11% [2].

There is much confusion about the projections of electric car production. It is difficult to predict the future, especially when there are such extreme projections bandied around. The following are such examples:

- In 2010, J. D. Powers, a most reliable car tracking company predicted that 7.3% of the car would be electric by 2020. However, according to the International Energy Agency (IEA), EVs sales only increased in 2020 by a 4.6% share of the new vehicle market, a 37% decrease in prediction.
- On August 5, 2021, Mathilde Carlier, a research expert transportation and logistics expert for STATISTA expects that there will be a cumulative 115 million vehicles in the global electric vehicle fleet by 2030, up from an estimated 8.5 million units in 2020 [3].
- CNBC reported that, according to the IEA, the number of electric cars, buses, vans, and heavy trucks on roads is expected to hit 145 million by 2030 [4].
- The IEA projects EV fleet development based on two scenarios the Stated Policies (STEP) scenario and the Sustainable development (SD), compatible with the Paris agreement. For 2030, depending on the scenarios, the IEA predicts 26 million

EVs under the STEP plan and 45 million cars under the SD plan [5].

- The global electric vehicle market size is projected to reach 34,756,000 by 2030, from an estimated 4,093,000 units in 2021 (Bloomberg). Clearly, the basis and documentation for such global predictions are not clear. It appears to be based on governmental projections. There is no information on whether the automotive industry has provided actual production data to support such numbers.

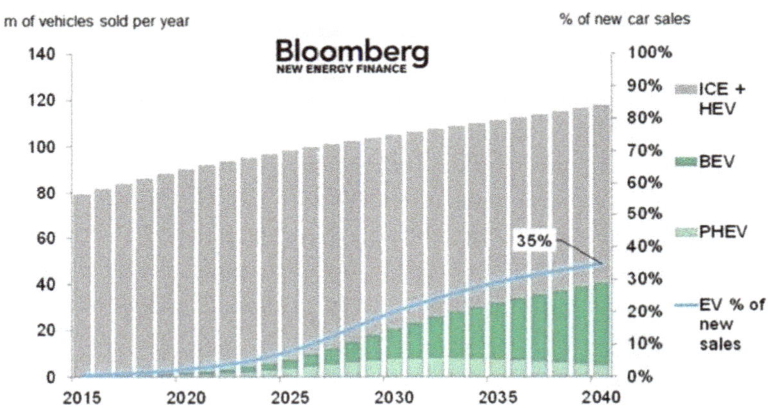

Figure 18.2 Bloomberg prediction of growth of various types of electric cars to 2030.

At the present time, China (4,710,000) and Europe (3,327,000) are the leaders in cumulative EV vehicle sales through 2020, followed by the United States (1,786,000). China has been aggressively pursuing long-term lithium raw material contracts with practically all lithium producers to satisfy its electric vehicle production plans. In the meanwhile, the Chinese CATL company is supplying lithium batteries to Tesla, Volkswagen, and the Chinese automaker GEELY, the Chinese automotive giant that sold 1.32 million units in 2020 [6].

There is also the issue of charging stations. President Biden has promised to deliver more than 500,000 charging stations as part of an infrastructure upgrade. This must mean that individual charging plugs will have to be installed, as there are only 168,000 gasoline stations in the United States [7].

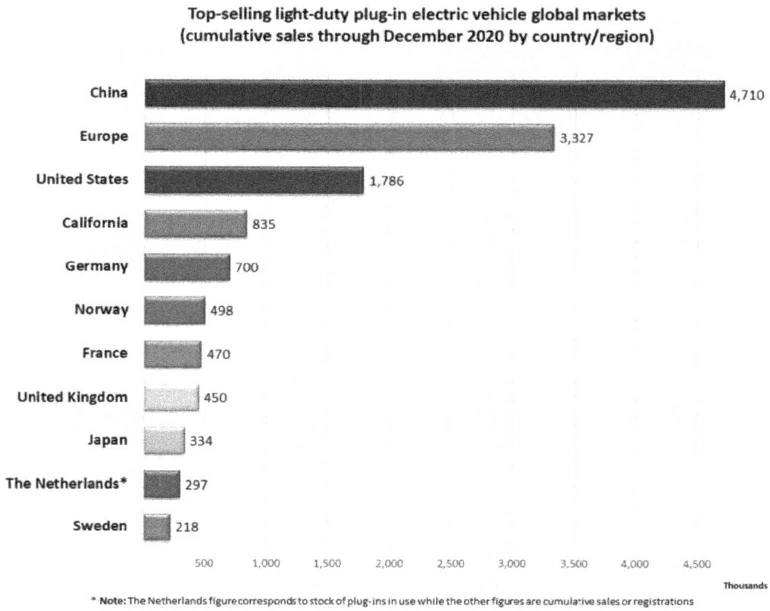

Figure 18.3 Cumulative production of light-duty electric various by country through 2020.

Under these uncertainties, questions are raised on the sufficiency of lithium to supply the ever-growing electrification of automobiles and storage facilities. A typical EV can have about 5,000 battery cells. Building from there, a single EV has roughly 10 kilograms or 22 pounds of lithium in it. A Tesla-type battery system requires 60 kg. Li for its 70-kWh battery system. A ton of lithium metal is enough to build about 90 electric cars. When all is said and done, building a million cars requires about 60,000 tonnes of LCEs. Thus, 28,000,000 cars in 2030 will require 1,680,000 tonnes of lithium carbonate (or hydroxide) and 2,400,000 tonnes of lithium carbonate in 2040 for the projected 40 million cars. The requirement will be 6 times higher for Tesla-type EV. Predictions suggest that there will be a global fleet of 145 million electric cars by 2030; this implies that 8,700,000 tonnes of lithium carbonate would have been produced to satisfy such a number.

From 2010 to 2020, 11 million electric cars have been produced. Statista reports a production of 564,000 tonnes of lithium for that period. Depending on the battery type (GM Volt or Tesla), 11 million cars would need anywhere from 264,000 to 660,000 MT of lithium, well within the production capacity of the industry. To extrapolate to 28,000,000 cars in 2030 and 40,000,000 cars in 2040 seems to put a real constrain on the supply side if only virgin lithium material is considered [8].

There is a clear disconnect between the lithium production capacity of the lithium industry and the predicted cumulative production of 145,000,000 electric vehicles. The EV predictions imply that an ever-increasing new virgin lithium material will need to be produced, which, based on the expected expansion capacity, questions the ability of the industry to meet such demand. However, Bloomberg NEF's latest lithium outlook shows enough lithium mining projects in the pipeline to meet demand. Additionally, the voices of the actual lithium producers merit attention. On August 5, 2021, Albemarle Corporation said that U. S. President Joe Biden's executive order aiming for half of the new vehicles sold by 2030 to be zero-emissions is already baked into automaker's plans and thus should not markedly change lithium demand [9].

Most analysts assume that the lithium required for the automotive and storage industries will be provided exclusively by primary lithium deposits and do not consider recycled lithium. The graph illustrates one of the scenarios showing that as the lithium available from recycling increases, the requirements from primary lithium deposits will diminish [10].

One of the factors that are not being considered in the discussion of the long-term lithium resource availability is recycled lithium. As more and more batteries reach their performance life (6–8 years), these batteries are and will need to be recycled as exceedingly more electric cars and storage units are produced. The DOE reports that 11 million tonnes of lithium batteries will reach the end of their life by 2030. The supply side will be alleviated because more and more companies have entered the battery recycling business:

- Glencore Recupyl SAS
- Australian Battery Recycling Initiative
- Umicore
- Redux Smart Battery Recycling
- Retrieve Technologies
- Battery Recycling Made Easy
- Nippon Recycle Center Corporation

- Raw Material Company Inc.
- The International Metals Reclamation Company Li-Cycle™
- American Manganese Inc.
- Global Battery Solution
- SAR Recycle
- Nippon Recycle Center Corporation

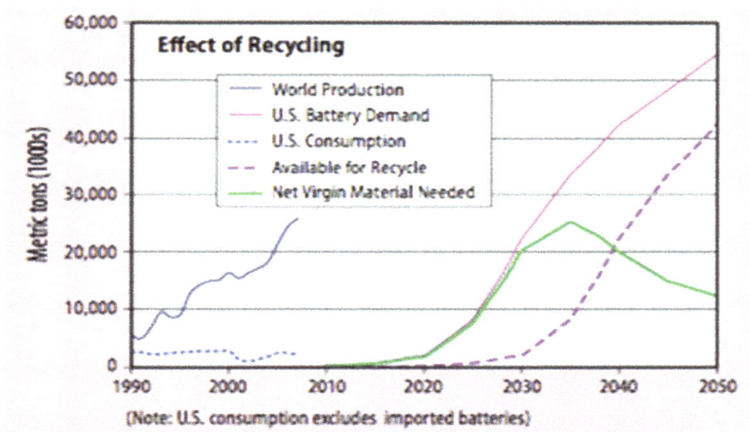

Figure 18.4 Effect of lithium batteries recycling on the predicted virgin demand for lithium.

American Manganese, with its patented RecycLiCo™ technology, has just announced that has produced a 99.99% purity lithium sulfate from recycled batteries [11]. Similarly, Rock Tech Lithium is a cleantech company with operations in Canada and Germany that aims to supply the automotive industry with high-quality lithium hydroxide "made in Germany." As early as 2024, the company intends to commission Europe's first lithium converter with a production capacity of 24,000 tonnes per year. This is equivalent to the volume needed to equip around 500,000 electric cars with lithium-ion batteries.

Companies, realizing the magnitude of the waste from discarded batteries, have already started recycling. Eric Melin of Circular Energy storage predicts that the United States will produce 80,000 MT and Europe 132,000 MT of Li-ion batteries. The amount of lithium available to produce new batteries depends on the percentage of recovery of various metals. Several laboratories are tackling the recovery of metals from batteries [12]. The following research groups are actively evaluating the recovery of metals from spent lithium batteries:

Established

- Umicore Valéas™ (Umicore, Bruxelles, Belgium)
- Retriev Technologies (Retriev)
- Recupyl Valibat (Recupyl)
- Akkuser
- Sumitomo–Sony (Sumitomo)

Emerging

- LithoRec
- Accurec
- Battery Resources
- Steven Loop: OnTo Technology (OnTo)
- Aalto University Process
- Others: BatRec, Inmetco, and Glencore

Rystad's analysis indicates that the battery lithium requirements are now 300,000 MT LCE per year (56,390 MT of contained Li), and that the capacity of 520,000 MT per year is enough to satisfy the growing market. But Rystad's prediction suggests a potential carbonate requirement of 2.8 million MT per year by 2028 [13], whereas using the 10 kg per battery scenario, only 1,680,000 tonnes of lithium carbonate will be required. Again, these numbers reflect virgin lithium, which is unrealistic considering the potential of recycling. Furthermore, these figures do not reflect the potential of alternative battery systems such as sodium and hydrogen (which are already being used in Europe). Therefore, it is not clear if lithium is the only option for electric cars and storage facilities. Certain events indicate that even after much research and development, lithium batteries appear to be unpredictable. On August 2021, GM recalled all 73,000 Chevy Bolts after having recalled 69,000 older models [14]. Volkswagen announced that it will produce electric cars only by 2035.

The uncertainty in realizing the very optimistic prediction for electric vehicles has recently been confirmed by a recent publication of the U.S. Energy Information Agency (IEA), which predicted that while the total vehicle miles would rise 0.9% per year, 78% of the vehicles sold in the U.S. will continue to run on gasoline for a quarter of a century. Furthermore it concludes that only 5% of the cars sold will be fully hybrids in 2040 [16]. This puts a significant damper on all the predicted exponential electric vehicle production and future investments into "green energy." In addition, several projections suggest that diesel will be an important fuel by 2050 (Fig. 18.5) [17].

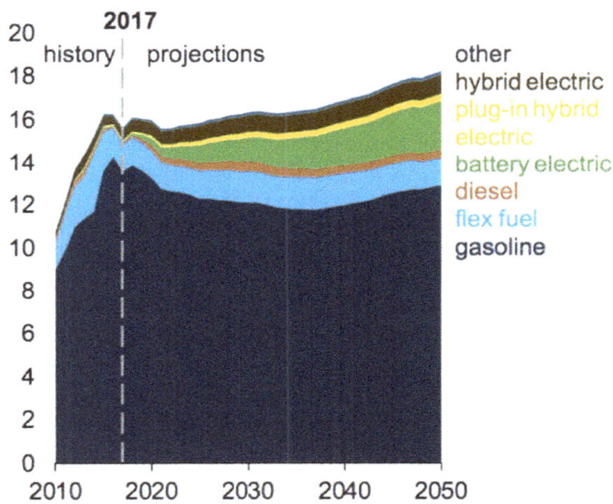

Figure 18.5 Light-duty vehicle sales by fuel type (millions of vehicles) [17].

In new developments, serious consideration is given to hydrogen-powered vehicles. Hydrogen, the most abundant element in the universe, is increasingly viewed as one way to slow the environmental impact of more than 1.2 billion vehicles on the road today.

In Germany, hydrogen-powered trains began operating in 2018, and Airbus is considering hydrogen fuel as well. Within three years, GM, the trucking firm J. B. Hunt plans to build fueling stations and run hydrogen trucks on U. S. highways. While it is claimed that hydrogen is environmentally friendly, it is produced from refineries and fertilizer manufacturing that use natural gas or coal.

Major automobile manufacturers are offering a limited but growing number of fuel cell electric vehicles (FCEVs) to the public in certain markets, in sync with what the developing infrastructure can support [15].

Technological advances are likely to improve the present situation.

References

1. Warren, I., 2021, *Techno-Economic Analysis of Lithium Extraction from Geothermal Brines*, No. NREL/TP-5700-79178. National Renewable Energy Lab. (NREL), Golden, CO (United States).
2. Roskill, 2018, Lithium: Global Industry, Markets and Outlook to 2027
3. Garside, M., 2022, Global projection of total lithium demand 2019–2030, www.statista.com
4. Frangoul, A., 2021, Global electric vehicle numbers set to hit 145 million by end of decade, IEA says, www.cnbc.com
5. IEA, 2021, Global EV sales by scenario, 2020–2030.

   https://www.iea.org/data-and-statistics/charts/global-ev-sales-by-scenario-2020-2030
6. Geely Auto 2020 sales reach 1.32 million units, www.global.geely.com
7. fueleconomy.gov
8. Garside, M., 2022, Global projection of total lithium demand 2019–2030, www.statista.com
9. Home, A., 2021, U.S. infrastructure bill targets critical minerals supply, www.reuters.com
10. Gaines, L. and Nelson, P., 2010, Lithium-ion batteries: Examining material demand and recycling issues. Paper presented at the TMS 2010 Annual Meeting and Exhibition, Seattle, Washington, February 2010.
11. The RecycLiCo patented process for sustainable lithium-ion battery recycling, www.americanmanganeseinc.com
12. www.mdpi.com
13. Lye, J., 2021, Soaring battery demand to spur raw material supply crunch, www.rystadenergy.com
14. GM recalls every Chevy Bolt ever made over faulty batteries, August 25, 2021, www.wired.com
15. Fuel Cells Electric Vehicles, Alternative Fuels Data Center, U.S. Department of Energy, www.afdc.energy.gov
16. U.S. Energy Information Agency, March 3, 2022, Early Release Overview, Annual Energy Outlook, www.eia.gov/outlook/aeo
17. Transportation 2050: More EV's, but Conventional Vehicles Will Still Dominate, EESI 2018.

Chapter 19

Future Projects

In the early days of lithium production individual companies invested in a few projects. Today, no trace is left of these early pioneers.

Now the bulk of the global lithium chemical and mineral production is controlled by a handful of companies, Albemarle, Soquimich, Livent, and Talison. Sizable mine activities continue in Zimbabwe. Smaller mines operate in Portugal, Brazil, Namibia, and Canada [1].

Recently, potential pegmatite deposits in the Democratic Republic of the Congo (Manono-Kitotolo) are undergoing extensive exploration activity.

Development rights on two spodumene deposits in Ukraine (Shevchenkisvke and Dobra), explored in the 1980's and 1990's and reported to contain 500,000 metric tons of lithium oxide, have been obtained by European Lithium, the operator of the Wolfsberg deposit in Austria.

The brines of the Salar de Uyuni in Bolivia hold very large lithium resources, but, to-date, various investors have not been successful in reaching an agreement with the government of Bolivia.

While the People's republic of China is investing heavily in brine development and research, it has not been able to achieve significant production. It satisfies its demand for lithium by securing long-term supply contract from existing brine and pegmatite operations.

The Lithium Legacy
Ihor Kunasz

ISBN 978-981-4968-31-7 (Hardcover), 978-1-003-37236-3 (eBook)
www.jennystanford.com

19.1 Jadar, Serbia

The underground deposit is in the Gender Valley, Serbia. Drilling by Rio Tinto has confirmed a reserve of 16.6 million tonnes grading 1.81% Li_2O and 13.4% B_2O_3. Contained in a mixed lithium boron mineral, it has been identified as a new mineral as jadarite by the Natural History Museum of London and the National Research Council of Canada. It was named after the Jadar Valley in Serbia. The mineral resource comprises 55.2 million MT of indicated resources at 1.68% Li_2O and 17.9% B_2O_3 with an additional 81.4 million MT of inferred resources grading at 1.84% Li_2O and 12.6% B_2O_3 for a total resource of 131.9 million MT of 1.78% Li_2O.

The first production from the mine is expected in 2026. Following the ramp-up to full production in 2029, the mine is expected to produce annually 58,000 MT of lithium carbonate and 160,000 MT of boric acid over the expected 40-year life of the mine, producing 2.3 million tonnes of lithium carbonate. This would make Rio Tinto one of the top 10 lithium producers in the world. The future of the project is, however, unclear as there is strong resistance. 110,000 petitions have been submitted against the project by the local authorities, leading to a decision by the President of Serbia to hold a referendum on the project [2].

19.2 Rhyolite Ridge, Nevada

The recent discovery of a deposit like Jadar located on the western side of the Silver Peak Range had been recently discovered in the Silver Peak Range west of the Albemarle lithium brine deposit. Considered the largest lithium boron sediment-hosted resource in North America, the deposit contains a total of 137 MT of ore grading 0.18% Li and 1.26% boron and a cutoff grade of 0.1% Li and 0.5% B. The company ioneer Ltd. has updated the total resource to 146 and a half million MT, including an ore reserve of 60 million MT for an estimated 26-year mine life. A recent announcement states that Ioneer, Ltd. has received a $700 million loan from the US Government to start operations to produce 24,000 tons lithium carbonate/hydroxide and 192,000 tons of boric acid per year.

19.3 Geothermal

Lithium production from geothermal brines has been touted for decades in the Imperial Valley of California. Cerro Prieto in Mexico has also been considered, although my evaluation and a sampling of the discarded brine showed up to 400 parts per million lithium, available volumes were not sufficient to warrant large-scale production.

Lithium production from the Imperial Valley geothermal brines field has eluded potential producers because a successful process has not been successful. Simbol Materials came the closest to a commercial operation treating 100 to 200 ppm Li. According to the USGS, Symbol planned to use a unique reverse osmosis process to extract high-purity lithium carbonate from the discharge brine eliminating the need for solar evaporation. Simbol again announced plans to produce 15,000 tonnes of lithium carbonate in 2015, but ran out of money and dropped the program. Three companies, Controlled Thermal Resources, EnergySource, and Berkshire Hathaway Renewables, are testing or scaling up pilot technologies at the Salton Sea. Controlled Thermal hopes to produce 17,000 tonnes of lithium carbonate by 2023 and then double the production by 2025 [3].

Today, although the California Energy Commission actually made the unsubstantiated claims that 600,000 tonnes of lithium carbonate could be produced, the chair of the Commission probably confused resources with production. While the technology proposed by Lilac has proven has been proven on a laboratory bench scale it has not addressed the type of plant required to accommodate a flow throughput of 6000 gallons per minute with 200 parts per million Li brine generated by the geothermal plant.

19.4 Great Salt Lake Chemicals

The Great Salt Lake brine is a residual brine created by the desiccation of the much larger Lake Bonneville. The lithium concentration varies between 20 and 60 ppm. Once the Silver Peak, Nevada, brine operation was initiated and produced lithium carbonate, Great Salt Lake Chemicals attempted to unsuccessfully produce lithium carbonate from the weak brines and decided to produce salt, potassium sulfate, and magnesium chloride. Lithium concentration within the ambient brine of the North arm of the Great Salt Lake ranges from 55 to 60 parts per million. After three years of solar evaporation process in the pond system, the lithium concentration in the brine after magnesium chloride production reaches up to 1000 to 1600 ppm Li, Compass explained. The lithium concentration in the IB Pond ranges from 205 to 318 ppm. As such, the company anticipates being well-positioned to serve the well wildly forecasted increase in domestic market demand for lithium [4]. Compass is targeting an annual production capacity of around 20,000 to 25,000 tonnes of lithium carbonate equivalent from an identified lithium brine resource of approximately 2.4 million MT lithium carbonate equivalent (LCE) of battery-grade lithium with up to 65% of the future production derived from brine that has already been extracted from the Great Salt Lake and is in varying stages of concentration within the existing pond system. In a separate annoucement, Cypress Minerals has reported a resource of 127,000 metric tons of lithium carbonate have been identified in the interstitial brine of pond 1B [4]. While the lithium resources accumulated in the existing pond system will support the announced production of lithium carbonate, and the fact that lithium is a by-product of potassium salt production, the ultimate sustainable production will be dependent on the steady-state potassium and magnesium production levels.

19.5 US Magnesium, Great Salt Lake, Utah

US Magnesium, LLC, is the largest producer of primary magnesium in North America, operating facilities on the Great Salt Lake where magnesium has been produced since 1972.

A recent announcement reports that the company has started the production of lithium carbonate to be sold to Sumitomo in Japan, Korea, and China [5].

In 2019, US Magnesium has contracted Wollam Construction to build a lithium carbonate plant from the 1.5 million tonnes of cell salt leftover from magnesium production. The facility, operated by US Magnesium since 2008, has now developed a way of extracting lithium from these seemingly insignificant leftover minerals (wollamconstrucction.com).

Bench-scale laboratory development and the operation of a pilot plant for six months led to the design of the commercial lithium carbonate plant with a nameplate capacity of 10,000 tonnes per year [6].

Table 19.1 Variations in brine chemical composition across the Great Salt Lake

	South arm shallow brines (d<20 feet)					South arm deep brines (d>20 feet)			North arm brines			
	March 1930[1]	April 1960[2]	1966[3]	1969[3]	1972[3]	1966[3]	1969[3]	1972[3]	December 1963[4]	1966	1969	1972
Cl^-	57.05	55.88	55.02	55.97	55.56	55.51	56.02	55.20	56.04	52.98	55.10	56.17
Mg^{++}	2.75	2.91	3.53	3.25	3.54	3.57	3.31	3.71	4.66	4.10	4.15	3.72
Ca^{++}	0.17	0.12	0.15	0.07	0.12	0.12	0.06	0.08	0.09	0.11	0.08	0.05
Na^+	32.90	32.71	31.54	30.70	30.36	30.95	30.32	30.27	29.08	31.59	29.34	29.20
K^+	1.61	1.71	2.35	2.45	2.64	2.26	2.34	2.69	2.75	2.79	2.80	2.95
SO_4^-	5.47	6.60	7.35	7.48	7.69	7.53	7.87	7.99	7.28	8.36	8.44	7.83
B	–	0.01	–	0.01	0.03	–	0.02	0.02	0.01	–	0.01	0.01
Br^-	–	–	0.04	0.05	0.05	0.04	0.05	0.03	–	0.05	0.06	0.05
Li^+	–	–	0.02	0.02	0.01	0.02	0.01	0.02	–	0.02	0.02	0.02
SiO_2	–	0.02	–	–	–	–	–	–	0.001			
Bicarbonate as carbonate CO_3^-	0.05	0.06							0.09			

[1]Most recent published prefill analysis (Hahl and Handy, 1969, p. 14).
[2]First published analysis of shallow (d<20 feet) south arm brines after completion of railroad fill (Hahl and Handy, 1969, p. 14).
[3]Median values of average analysis for each sampling date sum recomputed to 100 percent.
[4]First published analysis of north arm brines after completion of railroad fill (Hahl and Handy, 1964, p. 14).

19.6 Oil Field Brines

The brines of the Smackover Formation have long been known to hold significant lithium concentrations [7]. Albemarle and Chemtura produce bromine in Arkansas. After bromine recovery, the tail brine contains about 200 ppm Li. Several oil companies have also evaluated the possibility of recovering lithium and other chemicals from the separated brines after oil is pumped to the surface.

The state of Arkansas reports that the lithium resources are 3.2 million MT of LCE. Two companies are presently active:

- Standard Lithium
- Galvanic Energy

Standard Lithium in an agreement with the German company Lanxess has now announced plans to produce lithium from spent bromine brines. With the rights to the Lanxess extraction technology, the company states that it will be a global leader in Direct Lithium Extraction (DLE) technology. Because of the location of the brines in southern Oklahoma where evaporation is not feasible, Standard Lithium has opted to apply a different methodology using a solid ceramic adsorbent to extract lithium in a matter of hours rather than the months required by salar brines. The company report that they control 4.335 million MT of LCEs and plans to produce 20,900 MT after full operation for 25 years. The company also has an option to acquire some additional exploration and production rights southwest of their property where lithium is as high as 400 mg/l with the opportunity to annually produce 30,000 MT of lithium hydroxide, a preferred chemical for the production of lithium metal (standardlithium.com). It is not clear whether or not the process will be able to handle the 5.3 billion gallons of brine pumped annually.

Galvanic Energy is another entrant into the potential recovery of lithium from a 100,000-acre prospect adjoining Albemarle and Tetra properties. The company reports a resource of 4,000,000 tonnes LCE under their prospect (galvanicenergy.com). The company will use the same DLE technology proposed by Standard Lithium.

In Canada, E3 Metals corporation is proposing to use Livent's ion-exchange DLE process to treat brines from the Clearwater resources area in Alberta Canada. While the company reports a total LCE resource of 7 million tonnes, contained in the Leduc formation.

The principal target has a lithium concentration of 74.6 mg/l (E3metalscorp.com) low by any global standard and possibly uneconomic when compared to the Oklahoma projects of Standard Lithium and Galvanic Energy.

19.7 Clays

Clays have been touted as a source of lithium for over a decade. However, the recovery of lithium from such low-grade material has been elusive. Since the lithium is of such a low concentration, there will be a huge amount of waste, practically 100% of the mined material. While presently realistic, these and additional projects will advance the knowledge of lithium in clay deposits.

19.7.1 McDermitt, Nevada

Following the detailed 1978 work by Glanzman, Rytuba, and McCarthy [8], Lithium Americas was the first company to attempt the recovery of lithium from clays at the Thacker past deposit, reporting a reserve of 179,422,000 MT of 0.32% Li. Unfortunately, their attempt failed to extract lithium economically using sulfuric acid and instead produced a specialty organoclay—hectatone—supplied to the well drilling industry [9]. The Australian company Jindalee Resources Ltd. is the latest company to attempt to produce lithium from this deposit. Its evaluation resulted in an indicated and inferred resource of 1.3 billion tonnes grading 0.132% lithium at a cutoff grade of 0.1% lithium [10].

19.7.2 Clayton Valley, Nevada

Cypress Minerals explored the Tertiary Esmeralda formation exposed on the eastern side of the Albemarle brine operation. This Tertiary clay had been previously evaluated by the author during his doctoral thesis on the origin of lithium in Clayton Valley. His work identified hectorite—the rich lithium clay. Samples collected by Cyprus Minerals have returned lithium values ranging from 0.06–0.12% Li. Cypress has abandoned this project as it clearly would not provide the basis for an economic lithium-extraction process [11].

19.7.3 Sonora, Mexico

A third clay deposit has been investigated in the state of Sonora, Mexico. Bacanora/Sonora Lithium Ltd. published a measured and indicated mineral resource of 291 million tonnes, with an average of 0.325% Li. If successful with a roasting cycle followed by a sodium sulfate treatment to produce lithium sulfate, the company would produce 17,500 tonnes of lithium carbonate annually [12].

19.7.4 Arizona

Several companies are conducting a preliminary exploration of several Arizona clay occurrences:

- Hawkstone: Big Sandy Lithium
- Ameriwest Lithium: Thompson Valley hectorite deposit
- ZenithMinerals: Burro creek

19.7.5 St. Austell, China Clays, England

Mineralogical studies showed that nearly all of the lithium is contained in mica with a lithium concentration ranging from 0.5% to 2.5% Li. Theoretically, a large proportion of the mica could be extracted as a by-product of China clay production. Lithium-containing mica could be crushed and milled. Flotation of the milled material could yield a 2% lithium concentrate [13]. Efforts are underway to evaluate the recovery of lithium-bearing mica from the existing kaolin waste piles.

References

1. Jaskula, B. 2020, Lithium Mineral Commodity Summaries, United States Geological Survey, 98–99.
2. Thousands protest against Rio Tinto's lithium project in Serbia, December 6, 2021, www.mining.com
3. Wilson, J. and Rode E., May 13, 2022, Lithium Valley: A look at the major players near the Salton Sea seeking billions in funding, www.desertsun.com
4. Compass Minerals identifies approximately 2.4 million metric tons of

Sustainable Lithium Resource, July 13, 2021, www.compassminerals.com

5. Announcing the conclusion of an exclusive agent for lithium carbonate produced in the United States for sale in Japan, Korea and Japan, July 22, 2020.
6. usmagnesium.com
7. Collins, A. G., 1976, Lithium abundance in oil field waters, in *Lithium Resources and Requirements by the Year 2000*, Geological survey professional paper 1005.
8. Glanzman, R. K., McCarthy, Jr., J. H., Rytuba, J. J., 1978, "Lithium in the McDermitt caldera, Oregon, in Nevada, *Energy*, 347–353.
9. Lithium Americas' Hectatone Subsidiary Enters Into Strategic Alliance with TOLSA, www.lithiumamericas.com
10. Jindalee repositioning as a pure-play U.S. lithium company, www.jindalee.net
11. Cypress Development Corp.: Large lithium resource in Nevada, www.cypressdevelppmentcorp.com
12. Bacanora Lithium Project Overview, www.bacanoralithium.com
13. Hawkes, J. R., 1987, The lithium potential of the St. Austell granite, www.inis.iaea.org

Chapter 20

Lithium: The Future

While much information is available on lithium and this book has attempted to elucidate some of the interesting aspects of the third element, there are a number of issues that still need to be clarified when trying to predict the future of lithium in the coming decades.

- Do the predictions of electric car use accurate?
- Are the lithium resources adequate to support such predictions?
- Are the existing producers capable to supply the lithium for battery applications?
- What is the impact of the pollution by used lithium batteries on the environment?
- Will there be a sizable recycling industry to handle the used lithium batteries?
- Is lithium the only option for batteries?

Experts conclude that the electrification of vehicles has been largely due to the issue of climate change and environmental concerns. Governments are clearly moving toward restricting vehicle emissions, even requiring zero-emission, and municipalities are pushing for banning the internal combustion engine (ICE) vehicles. Thus, the highly optimistic predictions. Vehicle sales are driving automakers to invest in the space [1]. While China has been leading in research, development, and sales, Europe has now overtaken

The Lithium Legacy
Ihor Kunasz

ISBN 978-981-4968-31-7 (Hardcover), 978-1-003-37236-3 (eBook)
www.jennystanford.com

China. However, because of the large population, China is expected to lead. The success of EVs has been dues to the large subsidies in China and Europe.

In Chapter 18, Lithium Demand and the Electric Car, projections on the number of electric automobiles reach 23 million in 2030 and 48 million in 2040. Today's ICEs still make up 90% of the total number of vehicles sold. The EV market growth was, however, supported by important incentives, which are decreasing as sales have improved. The final price of these cars is still much higher than their ICE counterparts; numbers show that EVs are 37% more expensive than ICE vehicles in Europe, 9% in the United States, and 37% in China [2]. Higher penetration may depend, not on the political fiats, but rather on the purchase cost, when subsidies are no longer available. The adequacy of lithium resources is only part of the equation.

Global consumption of base lithium compounds in 2017, according to Roskill [3], was approximately 144,000 MT LCE, representing 68% of the total consumption of lithium compounds on a lithium carbonate equivalent (LCE) basis. Roskill forecasts that base lithium compound consumption is expected to grow to approximately 417,000 MT LCE by 2027, representing an 11% compounded annual growth rate and 47% of the total consumption of lithium compounds. According to Roskill, the global supply of all lithium compounds was 270,000 MT LCE in 2017, amply sufficient to satisfy the demand. Based on the projected increases in production by the major producers, the 417,000 metric tons LCE demand will be satisfied. The six producers, including Livent, Albemarle Corporation ("Albemarle," previously Rockwood Holdings, Inc.), Sociedad Quimica y Minera ("SQM"), Tianqi Lithium Corporation ("Tianqi"), Jiangxi Ganfeng Lithium Co. Ltd. ("Jiangxi Ganfeng Lithium") and Orocobre Limited ("Orocobre"), accounted for approximately 86% of the total 2017 supply. The Chinese production is basically from Australian spodumene concentrates. As discussed in prior chapters, demand for lithium has been predicated on the availability of virgin lithium. However, recycling has clearly become increasingly important because of the large environmental impact of disposing of the millions of lithium batteries discarded after their useful life. Therefore, the production requirements of virgin lithium will be significantly reduced.

However, two other metals are critical: cobalt and nickel.

The Democratic Republic of the Congo produces half of the world's cobalt: Glencore; 27,000 MT; Fleurette Group, 7600 MT; Gecamines, 4200 MT; with China Molybdenum at 16,500 MT and Vale at 5800 MT. Supply uncertainties have led automotive manufacturers to seek different supplies such as Australia and Morocco.

Finally, lithium is not the only element necessary to produce effective batteries. Research is being conducted on sodium and hydrogen systems.

20.1 Sodium-Based Batteries

Like lithium-ion batteries, the key components of sodium batteries consist of a cathode material, an anode material, a separator, and an electrolyte. Na^+ shuttle back and forth through the electrolyte and separator between the anode and cathode during the cycling process. While not applicable in rechargeable systems as is lithium, it is considered in energy storage systems where cost is more important than power. To advance the commercialization of sodium batteries, the U. S. Department of Energy's Advanced Research Projects Agency–Energy awarded Natron Energy, Santa Clara, California, $19.9 million as part of a new program to fast-track technologies [4].

20.2 Hydrogen-Based Systems

In 1845, English scientist Sir William Grove demonstrated that, given the right conditions and the addition of oxygen, hydrogen gas—also called H_2—can undergo a chemical reaction and produce a potent and non-polluting electrical punch.

Like all electric vehicles, fuel cell electric vehicles (FCEVs) use electricity to power an electric motor. In contrast to other electric vehicles, FCEVs produce electricity using a fuel cell powered by hydrogen, rather than drawing electricity from only a battery [5]. The car is equipped with a hydrogen fuel tank that charges a battery pack that provides energy to the car. Komatsu, the mining equipment producer, has announced that its heavy-duty mining haul trucks will run on hydrogen power by 2030 [6]. Airbus Industries has developed a turbofan jet design with a range of 2,000+ nautical miles, capable of operating transcontinentally and powered by a modified gas-turbine engine running on hydrogen, rather than jet fuel, through combustion [7].

So, what does the future hold? It was not long ago that lithium became a critical component of batteries. Today, two new candidates vie for that application. How many more elements will join the list? The answer is uncertain, but clearly human ingenuity will provide practical answers. As I asked in one of my publications, "*Quo Vadis, Litio*?" and the answer today is "Io" (I am doing just fine). Lithium is still the star.

References

1. www.investingnews.com
2. www.investing news.com
3. www.roskill.com
4. www.spectrum.ieee.org
5. afdc.energy.gov
6. asia.nikkei.com
7. www.airbus.com

Appendix

The 37 most abundant Li minerals and the number of occurrences:

Mineral Name and Formula	Number of Localities
Spodumene $LiAlSi_2O_6$	620
Elbaite $Na(Al_{1.5}Li_{1.5})Al_6(Si_6O_{18})(BO_3)_3(OH)_3OH$	506
Triphylite $LiFe_{2+}PO_4$	295
Amblygonite $LiAlPO_4F$	208
Lithiophorite $(Al,Li)Mn_{4+}O_2(OH)_2$	205
Montebrasite $LiAlPO_4(OH)$	185
Cookeite $(Al,Li)_3Al_2(Si,Al)_4O_{10}(OH)_8$	173
Lithiophilite $LiMn_2+PO_4$	137
Petalite $LiAlSi_4O_{10}$	111
Polylithionite $KLi_2AlSi_4O_{10}F_2$	86
Ferrisicklerite $Li_{1-x}(Fe_{3+},Mn_{2+})PO_4$	73
Sicklerite $Li_{1-x}(Mn_{2+},Fe_{3+})PO_4$	53
Neptunite $KNa_2LiFe_2Ti_2Si_8O_{24}$	50
Holmquistite &$Li_2(Mg_3Al_2)Si_8O_{22}(OH)_2$	37
Tainiolite $KLiMg_2Si_4O_{10}F_2$	37
Rossmanite &$(LiAl_2)Al_6(Si_6O_{18})(BO_3)_3(OH)_3OH$	35
Tavorite $LiFe_{3+}PO_4(OH)$	34
Eucryptite $LiAlSiO_4$	27
Trilithionite $KLi_{1.5}Al_{1.5}(Si_3Al)O_{10}F_2$	26

Fluor-elbaite $Na(Li_{1.5}Al_{1.5})Al_6(Si_6O_{18})(BO_3)_3(OH)_3F$	25
Manganoneptunite $KNa_2LiMn2Ti_2Si_8O_{24}$	25
Hectorite $Na_{0.3}(Mg,Li)_3Si_4O_{10}(F,OH)_2\_nH_2O$	17
Fluor-liddicoatite $Ca(Li_2Al)Al_6(Si_6O_{18})(BO_3)_3(OH)_3F$	15
Zabuyelite Li_2CO_3	14
Nambulite LiMn24 $Si_5O_{14}(OH)$	12
Sugilite KNa_2Fe3 $(Li_3Si_{12})O_{30}$	11
"Liddicoatite" $Ca(Li_2Al)Al_6(Si_6O_{18})(BO_3)_3(OH)_3(OH)$	10
Bityite $CaLiAl_2(Si_2BeAl)O_{10}(OH)_2$	8
Cryolithionite $Na_3Al_2(LiF_4)_3$	8
Griphite $Ca(Mn_{2+},Na,Li)_6Fe_{2+}Al_2(PO_4)_6(F,OH)_2$	8
Lithiophosphate Li_3PO_4	7
Sokolovaite $CsLi_2AlSi_4O_{10}F_2$ 7 32 87 12	7
Ephesite $NaLiAl_2(Si_2Al_2)O_{10}(OH)_2$ 33 5 42	6
Pezzottaite $CsLiBe_2Al_2Si_6O_{18}$ 6 34 76 52	6
Zektzerite $NaLiZrSi_6O_{15}$ 6 35 42 73	6
Masutomilite $KLiAlMn_{2+}(Si_3Al)O_{10}(F,OH)_2$ 5* 36 39 68	5
Norrishite KLiMn2 $Si_4O_{10}O_2$ 4* 37 53 44	4

Index

For Product Safety Concerns and Information please contact our EU
representative GPSR@taylorandfrancis.com
Taylor & Francis Verlag GmbH, Kaufingerstraße 24, 80331 München, Germany

www.ingramcontent.com/pod-product-compliance
Ingram Content Group UK Ltd.
Pitfield, Milton Keynes, MK11 3LW, UK
UKHW022159190926
475702UK00003B/107

* 9 7 8 9 8 1 4 9 6 8 3 1 7 *